"The emphasis on evolution and the psychological impact on the people involved in and participating in projects is a notable aspect of this work. It makes it clear that evolution does not occur without change within each of us.

The content will make potential readers reflect and help them understand that change is both for themselves and others who are part of the project; each with their own experiences. It encourages research and contains everything necessary for the reader to do so, improve their knowledge, and get involved. It contains clear evidence, lived and told by people who, ultimately, are agents of change.

A noteworthy chapter is 'Why this book,' which addresses and engages with the psychology underlying these changes."

Arturo McKeon
Past Division Technical Manager, Coca Cola Latin America,
River Plate Division

"In today's fast-changing and unpredictable world, managing change is one of the most critical skills for individuals and organizations. Change management is not only crucial for adaptation but also serves as the foundation for problem-solving, process improvement, and innovation, which often results in changes. This book offers a profound and practical understanding of this critical discipline, drawing from the author's extensive experience. It is an indispensable resource for managers, executives, and leaders who seek to navigate change effectively and drive meaningful progress."

Dr. Lars Sörqvist
Chair, International Academy for Quality
President and Partner, Sandholm Associates AB
Member of Advisory Committee, Shanghai Association for Quality (SAQ)

Projects Impacting Organizational Culture

Projects for continuous improvement are prevalent in today's business landscape, with transformations becoming increasingly common. However, individual and organizational resistance often hinders progress and inhibits results. That's where *Projects Impacting Organizational Culture: Evolution by Integrating People into the Equation* comes in. This essential guide offers a friendly and straightforward process to help implement and manage change effectively.

Whether you are re-adapting your organization's culture, using Blue Ocean strategy, or improving processes using methodologies like PDCA (Plan–Do–Check–Act) or DMAIC (Define, Measure, Analyze, Improve, and Control), this book provides a conceptually deep and practical process for application. It offers a unique perspective on the value of integrating social and technical disciplines based on core Quality Principles. This book presents a step-by-step process for conducting organizational diagnosis and interventions, combining theoretical knowledge with practical experience. Each step is accompanied by change management practices and tips, supported by relevant theory to ensure a solid understanding of the concepts discussed. Real case studies and examples are used to illustrate key points, integrating both social and technical approaches to change management.

This book is a valuable resource for professionals in design, manufacturing, project management, systems engineering, manufacturing management, reliability, usability, and quality specialists. It is also beneficial for graduate and senior undergraduate students, as well as instructors seeking to enhance organizational performance.

Best on Quality: Advancing Quality for Humanity
Series Editors: Elizabeth A. Cudney, Grace Brannan and Hiore Tsubaki

Each volume in the series is a collection of quality articles written by IAQ members, their collaborators, and invited experts. The books will explore quality concepts, practices, successes, and challenges facing the 21st century and beyond.

Each book will focus on bringing quality to humanity and expand from quality applications in the manufacturing and service industries to emphasizing and creating higher value for people and their quality of life with lesser use of resources. But more so, the books will talk about humanity as the main stakeholder for quality. Beyond satisfying customers, quality initiatives must fulfill the needs of society, their country, and the planet and be sustainable.

This new series, which will be a partnership between IAQ and CRC Press/Taylor and Francis, will include research, case studies, practical applications, and concepts on a variety of subject areas all related to quality for humanity.

If you are interested in writing or editing a book for the series or would like more information, please contact Cindy Carelli, cindy.carelli@taylorandfrancis.com.

Customer Centric-Design
Based on QFD Principles
David Menichelli and Glenn H. Mazur

Projects Impacting Organizational Culture
Evolution by Integrating People into the Equation
Edited by Raúl Molteni

For more information on this series, please visit: www.routledge.com/Best-on-Quality-Advancing-Quality-for-Humanity/book-series/IAQBOQ

Projects Impacting Organizational Culture

Evolution by Integrating People into the Equation

Edited by
Raúl Molteni

CRC Press
Taylor & Francis Group
Boca Raton London New York

CRC Press is an imprint of the
Taylor & Francis Group, an **informa** business

Designed cover image: Shutterstock

First edition published 2026
by CRC Press
2385 NW Executive Center Drive, Suite 320, Boca Raton FL 33431

and by CRC Press
4 Park Square, Milton Park, Abingdon, Oxon, OX14 4RN

CRC Press is an imprint of Taylor & Francis Group, LLC

ISBN: 978-1-032-89834-6 (hbk)
ISBN: 978-1-032-89757-8 (pbk)
ISBN: 978-1-003-54480-7 (ebk)

DOI: 10.1201/9781003544807

Typeset in Times
by Newgen Publishing UK

Contents

PART I   Introduction

PART II   The Basic Concepts

PART VII   Implementing and Sustaining the Change

PART VIII   And Now What

PART IX   Testimonies and Case Studies

Foreword

Although well understood, the observation that change is relentless and continuous neglects awareness that the rate of change is accelerating. Owing to social-technical advances and a growing world population – which has nearly tripled in my lifetime – rapid change is transpiring in practically all areas of life.

Within the world of work, leaders were once tasked with a heavy management role … keeping operations running. But now, leaders are primarily expected to have the acumen and skill to effect change, both incremental and transformative. A measure of sound management is still required from leaders, but the frontline worker is expected to perform to standard with less oversight. Indeed, the "hovering boss" is the brunt of much humor. Leaders at all levels are presumed to engage deeply in changing the organizations to ensure their survival.

Given the intensifying call for change leadership competence, what are the current learnings in this field? Raúl Molteni, a recognized global leader in change leadership, offers a fresh perspective which firmly promotes the human element of change. Equipped with extensive experience spanning the shop floor, middle management, C-suite, and the Board level, Molteni gifts a compelling narrative that combines theoretical concepts and practical insights to facilitate the path to successful change. At the core of his philosophy and experience is the belief that successful change is not just the implantation of new processes, technology, or structures; it involves reshaping the organizational culture. This book covers the methodology to effectively engage the human dimension of change to enact a capable organizational culture to make change efforts fruitful. Too often change initiatives focus almost entirely on the technical aspects and leave consideration of the human element behind. And when the concept of addressing organizational culture is introduced, unfortunately the tactics employed fall into the categories of "tell and/or sell" behaviors. Engagement is of a different nature than these.

Molteni explores the fundamental principles of change and speaks to the actions leaders must take to have people embrace a transformative change culture. How best to engage and inspire others is at the heart of his writings. His insights are well grounded in real-world experience as he draws from a wealth of successful change initiatives and provides case studies. He presents clear path for navigating the complexities of change initiatives, offering actionable solutions to engagement and culture development.

More than a straightforward guide, this book calls upon leaders to rethink their approach. It may be obvious, but as leaders ask for change, there is a requirement that they too must advance in their understanding and methods of creating and operating in a new culture. And, although practical tools and frameworks are provided, the underlying appeal is for leadership growth. Conventional approaches to change management have a less than stellar success rate. The call is for a more holistic and human-centered approach with respect to thinking, planning, and the execution of change.

As you read through the pages of this book, prepare to be challenged. Yes, you will gain knowledge. Molteni's insights are of immense value and his passion for change leadership infectious. But the wisdom required to create resilient, adaptive, and transformative cultures invites risks, risks embedded in altering leadership modeling from accomplishments in the past. Remember, the world has and will continue to progress at an accelerated pace. Keeping with the tried-and-true approaches of the past are in large part ineffective. The workplace and its participants have already changed. Will you?

Stephen Hacker
Transformation Systems International
Bend, Oregon, United States

Preface

This book is dedicated to all those who love quality, excellence, improvement, innovation, transformation, and evolution and see the changes they involve as a real adventure in which operational and human behavior are intertwined.

Methodologies such as Lean, Six Sigma, and Agile have been taught and applied according to best practices, and many changes in processes, products, and services have been addressed.

However, the results have not always been as expected. "People resist change" or "It wasn't for us" are some explanations that hide the real cause of failure. Resistances that seemed obvious when results didn't show up were initially ignored. A study by Prosci indicates that approximately 60% of projects that fail to meet expectations have social aspects that are not adequately addressed.

In practice, hard methodologies such as PDCA, 8D, A3, Lean, DMAIC, DMADV, Triz, and Agile incorporate activities designed to ensure adherence from those who are directly or indirectly impacted by the change. These are typically applied by those leading and executing the technical aspects of the change. These include Green, Black, and Master Black Belts, Lean Practitioners, Continuous Improvement or Operational Excellence Coordinators, and Scrum Masters, among other roles.

Meanwhile, methodologies such as change management, which draw on the approaches of Kotter, Prosci, or Changefirst, for instance, aim to address the social side of the change process more comprehensively. These methodologies are typically applied by professionals in human capital, human resources, culture, and change management.

What issues need to be addressed? In one scenario, these different perspectives result in the initiation of parallel projects with disparate objectives, schedules, and disconnected methodologies. In the other, the project is led by the "soft" or "hard" perspective without a solid methodology to complement it.

This book presents a change methodology that would integrate the importance of individual emotions, feelings, and opinions that arise with the technical approach. A methodology that recognizes the interdisciplinary nature of organizational work.

Change management and technical professionals will find solid fundamentals and common language for the analysis and proposals for improvement changes. A simple step-by-step process for planning and dealing with change management in a friendly and proven manner.

This book, not intended to be a "cooking recipe", helps you follow a friendly process to implement and manage change in a simple way. It aims to provide an original perspective on the value of creative and synergistic integration of Social and Technical disciplines based on the core concepts of Quality Principles.

It is hoped that readers will be inspired to continuously self-challenge, striving to enhance their abilities through experimentation, evaluation, learning, and ultimately the expansion and enrichment of their toolkit.

Raúl Molteni
President, International Academy for Quality
President and CEO, Molteni Consulting

About the Editor

Raúl Molteni is a partner and CEO of Molteni Consulting, Argentina. He leads transformation projects based on customer and employee experience and process management. He is currently the President of International Academy for Quality and has served as Chair for Quality in Governance Think Tank. He is a member of the Quality in Planet Earth Concerns and Quality in Logistics Think Tanks and has been on the Board of Directors from 2017 to 2018 for the American Society for Quality (ASQ). Molteni was awarded the Kaoru Ishikawa Medal by ASQ in 2021 for outstanding leadership and has had a positive impact on the human aspects of quality. He is a Certified Board Director for IGEP (Instituto de Gobernanza Empresarial y Pública) and an ASQ Six Sigma Black Belt.

Contributors

Luciana Barrera and her team assist organizations in their transformation, whether it involves digitalization, customer experience differentiation, cost optimization, process automation, or anything that brings them closer to their aspirations. Her experience has led her to hold positions such as Quality Director, Transformation Director, and CIO at Telefónica Argentina.

Hernán Eduardo Galdeano, more than 30 years at Ford Motor Co., has provided with the opportunity to rotate through technical, commercial, service, financial, and human resources areas. His undergraduate degree is in Mechanical Engineering, but he has completed graduate courses in Business Administration, Road Safety and Accident Studies, Negotiations, Six Sigma, and Process Leadership, among others. He is an Internal Auditor of the QMS under ISO 9001/2015, has trained in Ontological Coaching, and is a Visiting Professor at the Universidad Austral, Faculty of Engineering, and at the Advanced Diploma in the Automotive Industry in Argentina. As a consultant, he shares his knowledge and experiences with organizations on their journey to achieve their objectives.

Carlos Lucena, Special Contributor, has over 20 years of executive experience in HR at multinational companies—the Coca-Cola Company in Argentina and Uruguay—and more than 15 years of consulting experience in various business areas, including human resources, leadership, development and change, quality, communications, and knowledge management, helps organizations improve their competencies in organizational effectiveness, change management, culture, and communications.

Lorena María Gómez, with a Bachelor's Degree in Social Communication and being a Professional Ontological Coach and Certified Professional Coach, is dedicated to promoting, facilitating, and creating spaces for rapprochement, meeting points, and relationship channels to foster personal enrichment and encourage the discovery of meaningful learning.

Nancy Nouaimeh advocates for challenging the status quo and making excellence a habit through mindset shifts, continuous learning, and behavior transformation. As the founder of XcelliUm (UAE)—the first affiliate of the Shingo Institute in MEA—she supports organizations on their continuous improvement journeys through knowledge transfer and culture transformation. With over 25 years of experience across research, food, healthcare, education, construction, and service sectors, Nancy excels in delivering strategic operational objectives, leading cultural transformations, and driving service enhancements.

Acknowledgments

I am especially grateful to Carlos Lucena, who enthusiastically accompanied me from the initial idea to write this book through the whole process of discussing and imagining the chapters.

Equal gratitude goes to Lucila who has backed me up for more than 50 years, and as our priest asked us in our wedding, through thick and thin. With her smile, silence, talking, and sharing dreams. Moreover, mainly with her love and forgiveness.

To Pablo, Martín, Paula, and Clara for their silent company and support, for making me proud, for sharing dreams, and for allowing me to enjoy five beautiful, extraordinary, and motivating people enormously.

To all those who cared for and taught me selflessly throughout my life and career, particularly to my father and mother, Oscar Cecchi, Marcos Bertin, Jack Jones, Blanton Godfrey, Gregory Watson, and Liz Keim.

To all IAQ, for their trust in me for a key role and for their passion, genuine selflessness, dedication, and use of quality values to improve society's life.

To all the excellent and honest professionals with whom we shared projects, for their contributions and insights, and to clients who trusted us on their path to continuous improvement. In some way or the other, they all wrote a line in this book.

And to my friends, who helped me enjoy life by playing futbol, making wine, or sharing school, trips, and barbecues together.

Part I

Introduction

This section introduces the concept of change management. It begins by examining its reasons, purpose, and operation. Then, it presents the findings of a simple experiment to answer the question: "Is resistance to change significant when a technical project is well-planned?"

DOI: 10.1201/9781003544807-1

1 Introduction
Why This Book

Raúl Molteni

PROJECTS WITH AN IMPACT ON CULTURE

In business, projects can encompass a wide range of activities, from the maintenance of a plant to the vacation schedule of a production line, the transfer of an in-house call center to an outsourced one, or the closure of a silver mine. In each of these cases, the term "project" is used to describe the undertaking in question. Some projects are focused on creating or building something new, while others are focused on achieving a different status or level of performance. Still others are focused on creating something completely new and unique within the industry or field in which they operate.

Throughout my career, I have been involved in a variety of projects, including quality circle programs, process improvement using Lean and Six Sigma, product and service design, design and improvement of quality management systems, improving customer experience, improving employee experience, developing leadership competencies of managers and middle managers, improving aircraft maintenance timing and quality, installation of quality measurement for television programs, improving the timing of a warehouse operation, designing a theme park, closing a mine after its operation of over 50 years and developing communities that have been dependent on that mine for that time for post-closure economic sustainability, participating in a national program to create awareness of continuous improvement, plant closures, and mergers between companies with more than 5,000 employees, conducting strategic planning for companies with more than 50,000 employees, carrying out the strategic planning process for telecommunications companies, implementing the reengineering of a newspaper, and other similar initiatives.

Some are focused on a specific production or service unit, while others encompass the entire company. Some cover a large community, while still others address multiple communities. Some of our clients are in the production sector, including food, automotive, beverages, energy, entertainment, plastic products, packaging, fiber optics, foundries, metallurgy, theme parks, and plastic products. Others are in the services sector, including banks, call centers, consulting firms, hotels, insurance, logistics, telecommunications, mining, oil & gas, education—including schools and universities—health—including providers and hospitals—and entertainment—including TV channels and theme parks.

DOI: 10.1201/9781003544807-2

This experience revealed a common factor: culture is a significant inhibitor of transformation and evolution. Consequently, projects are delayed, take longer than expected, cost more than the planned budget, or the results are less than anticipated or not sustained.

With this book, I want to help prevent such consequences by sharing a simple but thorough methodology for approaching the human side of change, one that integrates technical and social mindsets.

The motivation is twofold:

- To ensure that most individuals adhere to the changes in a timely and comprehensive manner.
- To guarantee these individuals have the best possible experience throughout the change process.

It is insufficient to encourage people to "join our change." It is essential to demonstrate respect for the individuals whose work, relationships, and interactions with the organization are undergoing significant transformations.

2 Introducing Culture and Change

Raúl Molteni

LET'S BEGIN TALKING ABOUT TECHNOLOGY

Technology plays a pivotal role in today's world. However, an overreliance on technology can result in a race to incorporate the latest advancements, regardless of their actual value. It is not simply a matter of using apps or updating software. It is essential to consider the value we are adding for all stakeholders, including customers, users, employees, suppliers, and the broader community.

> *It is a matter of providing value to a stakeholder, whether that be a customer, user, employee, supplier, or member of the community.*

The concept of transformation, and the related terms we use to describe how organizations should approach the evolution we face, is not solely about technology. It is widely acknowledged that this is not the case. However, numerous assessments have consistently shown that the consequences of transformations are predominantly cultural in nature.

> *The consequences of transformations are typically cultural in nature.*

The evolution of technology offers valuable insights. Innovation does not originate from a significant enhancement in a specific field. Instead, it emerges from the integration of improvements across diverse fields, including artificial intelligence, mechatronics, sensors, and object recognition. We should take advantage of this example and apply the concept in our own business. During a transformation, it is essential to integrate not only efforts but also concepts and approaches, including technical and cultural.

> *During a transformation, it is essential to integrate not only efforts but also concepts and approaches, including technical and cultural.*

DOI: 10.1201/9781003544807-3

5

WHAT WE MEAN BY CULTURE

Without attempting an academic or universal definition for culture, let's consider it to be the beliefs, knowledge, experiences, and habits shared by the people that make up a certain organization—be it a business or a social group—and that will make those people react in a similar way to certain triggers—facts and expressions.

> *Culture: shared beliefs, knowledge, experiences and habits.*

The occurrence of an event in one cultural context may be perceived as ordinary and not produce any response from the individuals within that cultural group. The same fact could cause a strong response in another context. This phenomenon is also observed within the same organization, where subcultures may emerge that can hinder the effectiveness of change processes. As we will see, understanding each of these cultures and subcultures is essential for successfully implementing significant changes.

PROJECTS DRIVE CHANGE AND CHANGE IMPACTS CULTURE

In projects involving standardization, improvement, digitalization, evolution, and transformation, changes of varying types are anticipated. For example:

- *Customer relations.* Customer experience is a primary focus for most projects today, if not the primary focus. From our own experience, we know that customers often require time to adapt to new processes and solutions, and that the desired customer experience is not always fully considered. On occasion, modifications prove beneficial to some while introducing complications for others.
- *Processes, procedures, and working conditions.* In some cases, it may be necessary to alter established habits such as treating customers differently or moving from operating with large batches to smaller ones. Even in instances where a change streamlines processes, individuals are frequently dissatisfied with the alteration or exhibit unintended behaviors and habits.
- *The number of personnel* assigned to processes or activities. Drive reactions not only from those left behind, but also from people who imagine being in the same situation in the near future.
- *Organizational structure.* Certain changes affect the levels of responsibility and span of control. Opportunities for growth or access to positions of greater responsibility disappear, and some are threatened with positions of lesser responsibility.
- *The skills required* for a position.
- *The evaluation of personnel and performance.*

- *Relationship with third parties and the environment.* Changes may also affect the way in which people in the organization must relate to suppliers, to society, or to the environment.

The projects aim to substantially change the habits, behaviors, rules of relationships, and even the beliefs of those who, at some point in the past, were selected, recognized, and promoted by the same organization.

> *The projects aim to substantially change habits, behaviors, relationship rules, and even beliefs. How can we expect everyone to welcome the project and its changes with open arms?*

CULTURAL IMPACT INFLUENCES PROJECT OUTCOMES

The reaction of people—directors, managers, middle managers, employees, operators, customers, suppliers, and community members—to changes also has a boomerang effect on projects. The sum of the reactions leads to changes that are not accepted, delayed, partially fulfilled, or fulfilled only for a few days or months. The results may be null or below expectations and are achieved with resources and time far beyond what was planned.

> *The reaction of people—directors, managers, middle managers, employees, operators, customers, suppliers, society—to change has a boomerang effect on projects.*

CULTURE IS A KEYWORD

Culture is not as intangible as you might think. Culture is why people react and why they don't, what they say and what they don't say, and how they behave when a trigger occurs. For projects, it is what people think about it, what they believe, how much they trust in its purpose and potential benefits, how much they will do and say in favor of it, and how much they will be willing and able to sustain change.

> *Culture is not as intangible as you might think. Culture is why people react and why they don't, what they say and don't say, and how they behave when a trigger occurs.*

To be sound, a project must address both the technical and social aspects of change methodically. There are three key issues to consider:

- *Understand who the stakeholders are.* See her, see him, or see it. Not just listen to them. It's being them. Walking in their shoes. Feeling as they feel. Experiencing as they experience.
- *Understand his/her expectations and needs from his/her perspective.* Both explicit and implicit. Not what our resources allow us to do, or what our knowledge tells us we can do, but what we should do.
- *Understand all the interactions and moments of truth* our stakeholder has with the project. It's his/her perception, his/her evaluation, and his/her emotions that count.

> *It is essential to consider both the technical and social aspects of the change in question, employing a suitable methodology.*

As John Guaspari wrote in his book *I Know It When I See It,*[1] more than 40 years ago, and as Steve Jobs clearly stated, you will know if you are meeting those expectations and needs when they can touch, interact, use, and feel whatever it is you are delivering.

Here's the catch. To get them to interact, you have to act first, and analyze and discuss results with data. Data that comes from experimentation. Data that comes from short and fast cycles of listening, understanding, creating, experimenting, and learning with your customer, user, employee, or whoever the stakeholder is.

> *The data is generated through a series of brief and rapid cycles involving listening, comprehension, creation, experimentation, and learning with the customer, user, employee, or other relevant stakeholders.*

We need a culture that aligns with our aspirations. Where results drive learning, not punishment. We need a culture that matches our project. We can't sustain a project that ignores the culture.

> *We need a culture consistent with our project. We can't sustain a project that ignores the culture.*

Experience, seniority, and knowledge need to be re-thought and re-signified. In this ever-changing environment, what we will be able to do and how we will think in the future is more important than what we have already done. As Bob Gurr—"Original Imagineer"[2]— stated, "I am happy I did not know it was impossible."

> *We can't expect continuous improvement, innovation, experimentation, learning, and rapid response in a culture where an undesirable outcome of an experiment is considered a failure.*

DOES CULTURE IMPACT PROJECTS THAT STRONGLY?

Even if this resistance to change exists, does it really impact the quality, cost, and schedule of projects?

I have been conducting a kind of "experiment" for more than 20 years now.[3] More than 4,200 people from different countries, organizations, functions, ages, and experiences have participated. Groups of white-collar and blue-collar workers. Including boards of directors.

During the Lean and Six Sigma courses—with excellent participation and qualification of the participants, by the way—I would suddenly stop the discussion and give each group the same instructions: "Please count how many letters N there are in the paragraph on the page I'm going to give you. You will find the instructions on this page. It is a very simple task. Even though it is very simple, you have all the time you need." I gave them a paragraph with 46 words and 255 letters.

Questions for you: Do you expect errors in counting? If so, do you expect the number of errors to vary according to the level or function of the group? Do you expect the number of errors to vary according to the background or sector of the organization?

Well, these are the results. Typical results. From all groups. There is no difference by level, organization, industry, or country: always a majority of results are between 16 and 22, with a few results between 0 and 3. Occasionally, higher numbers could go as high as 247 (see Figure 2.1).

At the very beginning, I tried to show them that the final 100% inspection was not as effective as they had thought. The results were so "strange" that I was confused. I had been working or training with these groups for hours. There was a very good rapport and understanding between all of us. Questions from the group were common

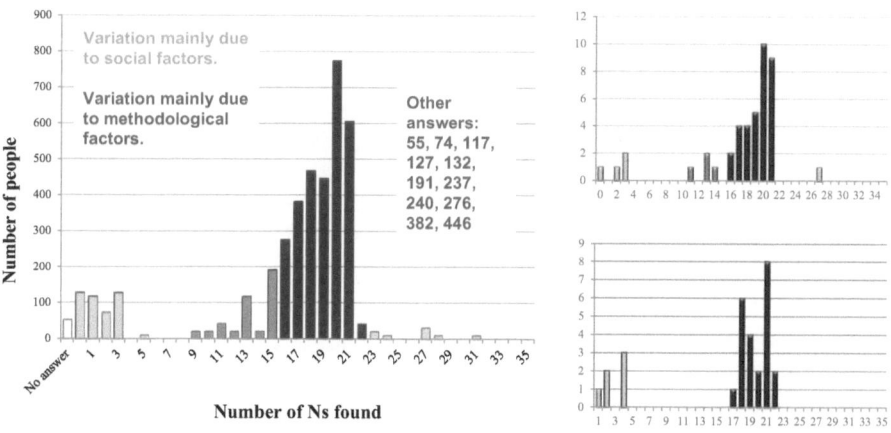

FIGURE 2.1 The distribution of N-letter count results by different groups and the consolidation of numerous groups. The group of bars at the center shows the results from those who counted the letters "N" from the named paragraph. The lower and higher numbers on the left and right show the results from those expecting a trap.

practice. But when the instructions were given, I knew there were doubts, but no questions were asked. Listening to the participants was enlightening; their past experiences made them react in a certain way.

Experiences they had had in the past made them react in a certain way.

To understand the results, 17–23 are the results of those who counted the number of letters N in the paragraph. Their mind told them: "Follow the instructions."

From 24 to 25 were the results of those who also counted within the said paragraph, but whose mind said: "My experience tells me that this is not so easy; just in case I add more N to the result." From 9 to 16 were the results of those who also counted within the said paragraph, but their mind told them: "Follow the instructions, but it is not worth it." From 0 to 5 were the numbers counted by those whose mind said: "This is a trap." So they found a trap: "The paragraph must be the next one or two sentences, that's it" (part of the instructions).

Numbers like 43 or 55 came from those who saw the letter "U" as a small "N" turned upside down, and scores like 237, 240, or 241 came from those who counted the N letters from the entire page.

Only groups of students from the Henry Ford High School in Buenos Aires—who have no work experience yet—responded with a different pattern.

So, what result should I have trusted if I didn't know how many letters there were?

	With Work Experience	Without Work Experience
Results below 17 and above 23	45.8%	0%
Results between 17 and 23	54.2%	100%

Even a simple change can lead to an unintended or unreliable result. This shows that what people are used to seeing and living in an organization triggers their behavior.

Even though an improvement or a new task may be seen as very simple and logical, the results may be severely affected by their assimilated habits. Even though the purpose of a change may be seen as positive, past experiences may tell people that they will face great risk and frustration and that it's better not to engage fully.

Your organization may not have the same cultural biases as those hundreds of groups and thousands of people. But it may have nonverbal instructions that "bosses are always right, so their voices should never be challenged." Or that "information is key, so don't share it openly." Or that "you have to look out for yourself, so teamwork is not something to strive for." Or "This is just a new fad; you just need to wait for the new CEO's fad."

There is no such thing as easy change if the culture is not ready.

So, when a change is implemented, you should expect people to react in some learned pattern. That means variation, delay, and nonacceptance of the change. No matter what it is, if you want people to be fully engaged—continuous improvement, innovation, evolution, transformation, more creative ideas, everyone being agile, thinking about their customers, quality, and teamwork—you need to understand how the culture will make people react. You may think it is not necessary. At best, you will go over budget, behind schedule, or deliver less than expected.

So, when a change is implemented, you should expect people to react in some learned pattern. This means variation, delay, and nonacceptance of the change.

NOTES

1 John Guaspari, *I Know It When I See It*. American Management Association, 1985.
2 Bob Gurr, "Original Imagineer." *The Imagineering Way*. Disney Editions, 2003.
3 Raul Molteni, *Change Management Is Not a Fad, It's a Fact-Based Need*. Sandholm Institute, 2022.

Part II

The Basic Concepts

This section addresses the fundamental concepts. These will be common to all stages of the design, planning, and implementation of projects that integrate the core technical and operational aspects with the social aspects, which are typically addressed within the context of change management.

It is crucial to have a unified understanding of the psychology of change, project coordination and integration, continuous learning, continuous communication, and owner, director, and management support for projects that impact business results before project planning begins. Such considerations are relevant at the outset of the project and throughout its duration.

DOI: 10.1201/9781003544807-4

3 The Psychology of Change

Raúl Molteni

THE BRAIN'S OPERATING PRINCIPLE

As neuroscientist Evan Gordon notes, "The fundamental organizing principle of the brain is the risk-and-reward response. Five times a second, at an unconscious level, your brain is scanning the environment around you and asking itself: Is it safe here? Or is it dangerous?"

When the brain is in a state of security, it is capable of functioning at its most sophisticated level. When it identifies potential risks and it perceives threats—as with any potential danger, such as coming across a tiger in the jungle—the brain triggers the protection mode, prompting a tendency to move away from the source of the threat.

A project represents a danger. The change promoter attempts to explain the logic and benefits of the project, but the brain of the interlocutor is tuned to "escape" from the danger. In other words, *when the brain is unsure of the situation, the default mode of response is to consider it unsafe and protect itself by retreating.*

It is crucial to design and create an environment for change that understands how our brain functions. When a change is imposed from a position of perceived superiority, with a logic without consideration of the receiver, information is withheld and managed with technical jargon, favors the action of a part of our brain—the Reptilian— and it reacts in survival mode. When treated with respect and understanding, another part of our brain—the Neocortex—is engaged and better positioned to objectively analyze and understand the situation: "Does the project align with my interests and goals?" and "Does it offer a positive and beneficial outcome?"

THE IMPACT OF CHANGE

No two changes are the same. Never mind the type of project, such as Six Sigma, Lean, merger, closure, leadership, or agility, the organization in which the project is developed is a key factor influencing its outcome. Previous experiences and culture all contribute to the unique nature of each project and its consequences.

To ensure success, projects must be designed considering three main factors:

- The changes proposed by the project.

DOI: 10.1201/9781003544807-5

- The technical structure of the organization, including processes, facilities, products, and services.
- The organization's culture, including values, beliefs, previous experiences, and decision-making processes.

"Culture; culture is key" Shahzad Ansari.[1]

THE THREE PHASES OF THE CHANGE PROCESS

One way to conceptualize the change process is through these three stages:

- The current situation is identified as unsatisfactory, prompting the question: Why change?
- Creation of the vision of the desired situation, prompting the question: What is the change for?
- Transition design, prompting the question: How to move from one situation to the other?

DISCONTENT WITH THE STATUS QUO

The current situation is perceived as unsatisfactory and is causing distress. There is a need to look for and try new ways of working. This is the basis behind the proposed change. If there is no clear and agreed-upon basis for dissatisfaction with the current situation, there is no motivation to embark on a project with a high impact on the business. If you find an explanation justifying why things are as they are when inviting someone to a change, then you are facing someone whose buy-in is far from being real. The use of data provides clarity and objectivity for analyzing the initial situation.

VISION OF A BETTER FUTURE

Initiating change requires a clear objective, a "What for", not merely a "Why." The "Why" is about identifying areas of dissatisfaction with the current situation. However, change also requires a clear, motivational, and beneficial vision of the future. The "What for" must be based on two key elements: *data and a future perspective.*

By necessity, by referencing or recommending, by comparison, by our own previous experiences or those of others, a vision of a future state that surpasses the current one emerges. This involves envisioning the desired future state and understanding the challenges and complexities involved in transitioning from the current situation to that desired state.

THE PATH, THE TRANSITION

If the path from the present to the vision is not defined, the change will remain a theoretical discussion. *Action is initiated when a program, method, or process demonstrates the approach to be taken.* A national quality award, the incorporation of an ERP, Design Thinking, Lean, Six Sigma, Agile, or a path to an Experience Centric Organization[2] are just some of the potential sources of inspiration for a path.

It is at this stage that many of the failures originate: a popular trend is embraced rather than a taking systematic approach that can support and facilitate growth. Six Sigma was adopted following General Electric and Motorola. Lean was implemented in order to emulate Toyota's approach, or alternatively, Agile was adopted with a view to replicating Spotify's methodology. Finding the way forward should involve the design of a customized strategy and plans that are aligned with the specific reality, ambitions, and culture of the organization.

> *Finding the way forward should involve the design of a customized strategy and plans that are aligned with the specific reality, ambitions, and culture of the organization.*

MODELS FOR CHANGE

There are several models that I consider to be of great value.

KUBLER-ROSS MODEL

Kübler-Ross researched the stages through which a complex experience that encompasses physiological, cognitive, and other factors takes place. The Model outlines five aspects:

- Denial, Anger, Bargaining, Depression, and Acceptance.

These are the stages that people find themselves in when facing very difficult situations as a defense mechanism against the problem.

As Elisabeth Kübler-Ross points out, a person does not necessarily have to go through all of them, they may even feel several stages at the same time. Their duration can also vary.

ADKAR MODEL, FROM PROSCI

This model "enables leaders and change management teams to focus their activities on what will drive individual change and therefore achieve organizational result."[3] It is based on the understanding that *organizational change can occur only when individuals change*.

It proposes five elements to achieve sustainable change:

- Awareness, of the need for change.
- Desire, to participate and support the change.
- Knowledge, on how to change.
- Ability, to implement required skills and behaviors.
- Reinforcement, to sustain the change.

EIGHT-STEP STRATEGY FOR CHANGE MANAGEMENT BY JOHN KOTTER

Kotter, in his book *Leading Change*,[4] outlines eight steps to achieve buy-in for change. Each of these steps addresses one of the fundamental flaws it encounters in transformation efforts.

- Create a sense of urgency.
- Build a guiding coalition.
- Form a strategic vision.
- Enlist a volunteer army.
- Enable action by removing barriers.
- Generate short-term wins.
- Sustain acceleration
- Institute change.

The first four help to unfreeze the status quo, the fifth, sixth and seventh propose new practices and the last step links the change to the culture.

THE ORIGIN OF RESISTANCE

The mere appearance of a change immediately takes us to our beliefs, values, and past experiences. To the knowledge—or lack of it—we have about the aspects or components of the change, and to what those we consider—bosses, colleagues, friends, parents—have to say. Then we shoot ourselves messages; we tell ourselves things about the change—for or against, depending on those messages—and we create emotions—positive or negative.

Now change is not something that others talk about. There is something of us in it, and so we make judgments. We give our opinion of this change based on our individualistic messages and emotions. At this point we have decided to rely on our judgments that lead us to a position of "it works for us" or "it doesn't work for us."

Resistance arises. The proponents of the change, with their judgments, will see certain positive aspects of the change. Others will see the negative aspects. There is a gap between the incomprehension of those who are confused because others do not see the "very logical" benefits of the project and those who do not understand how the promoters do not see the negative aspects.

This is what change management is all about: bridging the gap.

RESISTANCE EMERGES BEFORE CHANGES ARE IMPLEMENTED

People's emotions, and therefore their opinions and reactions to change, begin the moment they become aware of "something" about the change. That something could be a plan, an informal communication, a rumor or a question in a meeting that raises suspicions. It is often difficult to get full buy-in for change from those who are negatively affected by it in some way. But it is even harder when they have formed an opinion that threatens their comfort zone. It will be important, during the planning and implementation, to clearly understand what this change means to the people affected by it. Both in its operational aspects—those related to their job—and in its emotional aspects—what it creates for them.

My experience invites us to start the change management before technical and operational management—improvement, agility, innovation, transformation. Once people hear what the change will be like, be it data or inferences, it is too late to install measures to prevent resistance.

> *Change management must be anticipated at the start of the project.*

FACTORS GENERATING RESISTANCE TO CHANGE

Resistance to change is a normal and natural response to change. There are two type of factors that cause us to reject or resist change:

- *The nature and function of the person we are.* The economic situation and security, the opportunities for development, age, gender, the place and environment in which we live, our state of health, our aspirations, our view of our competencies, our future possibilities, and our capacity for development are examples.
- *External factors, factors that are alien to us.* How much we know about the change, who is proposing it, how much we trust that person, how much we have been involved, and how much we will be involved in its development.

> *The key question is not how to avoid resistance to change. The question is: how much resistance can we anticipate and avoid?*

INDIVIDUAL RESISTANCE

When we talk about change, we usually find individual resistance. This resistance comes from the people whose values, beliefs, and behaviors are being asked to change, or for whom the environment in which they work is being changed.
Rick Maurerv[5] believes that nearly 70% of significant changes will fail due to:

- "I don't get it." People may not understand what the change is about or what it is trying to accomplish, so they reject the change before they even try to understand it.
- "I don't like it." They understand, but for some reason they do not like it. Emotional barriers, personal interests, or their plans may not be favored by the change.

- "I don't like you." If employees do not trust the judgment of those proposing and promoting the change, they are likely to resist it.

Factors and Consequences that Create Individual Resistance

- *Uncertainty*: not knowing what will happen and how it will affect us.
- *Loss of control*: over what we do, over our future, over our possibilities for development.
- *Everything being different*: it changes and makes us uncomfortable. The environment, how and with whom we interact, what we do, or for whom we do it.
- *Increased workload*: a greater number of tasks or a more complex view of them.
- *Lack of confidence in our ability*: about whether we will be able to meet what appear to be new requirements and responsibilities.
- *Work environment*: change of physical location or space where we work.
- *Perception of being accused*: perceiving that change makes us appear to be wrong or doing something wrong.
- *Loss of autonomy*: we are asked to share or have someone else intervene when we used to do it alone.
- *The ghosts of the past experiences*: encountering a change that has been previously attempted and was unsuccessful, at least in our view.
- *Wave effects*: same change has not been favorable for other sector, unit, or organization.
- *Real threat*: there is a clear possibility that we will be harmed, either by losing growth possibilities or by being left out of the organization.
- *Different work environments*: some employees may have become accustomed to virtual work and may find it challenging to return to face-to-face work.

ORGANIZATIONAL RESISTANCE

The immune system of the organization is much more powerful than individual resistance. Blaming individuals is simplistic. Organizations also create resistance.

What is "organizational resistance"? Any organizational values, beliefs, policies, technological infrastructure, processes, procedures, practices, and metrics used that are not aligned with the intended change. Relationships with unions, society, and legal barriers. How decisions are made, is power centralized or distributed? Why and how are people hired, recognized, rewarded, and promoted? What do people consider and think about making mistakes and experimenting? Have past changes been successful? Are projects well planned, or is "just do it" the way things are done?

As Kris Østergaard states in *How Big Companies Can Simultaneously Run and Reinvent Their Businesses*,[6] "To create a strong innovation culture, an organization needs to thoroughly understand its immune systems. These are the mechanisms that protect the organization and operate around the clock to keep it healthy and stable, just as the body's immune system operates to keep the body healthy and stable."

> *Blaming individuals is too simplistic. Watch out for the organization's resistance.*

As we will see, this is a main target area for owners, board members, and CXOs. Their job is to address the organizational resistance.

THE CHANGE GENERATES 2-6-2

It was the mid-1990s, and it was during one of the public presentations in which the Malcolm Baldrige Award winners described how they had achieved this level of recognition. Three senior executives from three winning companies and a coordinator sat on a stage while about 200 people listened intently.

The moderator's final question was, "What have you learned?" The answer from a senior executive, whose company was renowned for its quality standards, was: "We learned 2-6-2." Seconds later, he continued, "We learned that in any change, there are two people who immediately embrace and even try to accelerate the change; they are enthusiastic, they are proactive. There are two others who strongly oppose the change: "it's not for us," "we've tried it before, and it didn't work." And that there are six people who are watching management's reaction to take one position or the other."

It is impossible for everyone to agree on a change that will have a positive impact on the organization's results, even if this defies your logic. To achieve the critical mass necessary to move forward, the position of management is key. They must differentiate between those who get on board and those who do not.

> *What happens when it does not happen—not adherence—and what happens when it does happen—adherence—The difference must be clearly visible to all.*

TIME REQUIRED FOR CHANGES TO BE OPERATIONAL

Jeffrey Lickert and David Meier[7] state: "One of the problems with the Kaizen event approach is that one week is not enough to change a culture." Changing culture takes time, and those who don't understand what it is and what it means can't understand that it takes time to change it.

> *Changing culture takes time, and those who don't understand what it is and what it means can't understand that it takes time to change it.*

Anxiety about results coming from top management is something to be aware of because it is a usual signal of top management not "seeing themselves" as a role model but expecting internal and external consultants to make all other people change.

WHY WE TALK ABOUT CHANGE MANAGEMENT

As previously stated, we have a range of technical methodologies at our disposal. We propose more agile methodologies designed to solve problems, improve processes, and facilitate the redesign of processes, products, and services. Change management is the equivalent methodology specifically designed to manage the social aspects of change.

> *Our objective is to encourage more people to adopt the required behaviors for the project in a shorter timeframe, while also fostering the development of essential habits and routines.*

CHANGE REQUIRES STANDARDIZATION—EMBEDDING IN CULTURE

Buy-in of change requires efforts to sustain it. It is not that once we have gained awareness of the need for change, we will never have doubts about it again. It is not that once we have achieved the commitment to participate, and the knowledge to act as the project requires, we will never have to work on these aspects again. It is a continuous work until it is embodied in the culture.

PUT YOURSELF IN OTHERS' SHOES, LISTEN, UNDERSTAND, AND LEARN FROM OTHERS

Ginnie Gallo[8] suggests a very interesting exercise. Write your full name with your normal writing hand and then with the other hand. And look at the result. How easily can you write it and then how much effort does it take to do it with the nondominant hand. Can you notice how it feels? How long does it take? How much concentration does it need? Use this exercise as a quick reminder to realize how people in the organization will feel about the change you want to make. Then forget thinking only about how to get them to change. Think how you can make it easy and less painful for them to adopt to the change. The ability to understand likes and dislikes will give you the ability to advocate and defend your project. You need to engage with the people you are trying to buy into the change to understand what their real needs are.

> *Forget thinking only about how to make them change. Think how to make it easy and less painful for them.*

NOTES

1 www.st-edmunds.cam.ac.uk/people/professor-shahzad-ansari/
2 Simon Clatworthy, *The Experience-Centric Organization, How to Win Through Customer Experience*. O'Reilly, 2019.

3 Jeffrey M. Hiatt, *ADKAR, How to Implement Successful Change in Our Personal Lives and Professional Careers*. Prosci Research, 1967.
4 John P. Kotter, *Leading Change*. Harvard Business Review Press, 2012 and www.kotterinc.com/methodology/8-steps/.
5 whatfix.com/blog/10-change-management-models/.
6 Kris Østergaard, *How Big Companies Can Simultaneously Run and Reinvent Their Businesses*. Singularity Hub, 2019.
7 Jeffrey Liker and David Meier, *The Toyota Way Fieldbook, a Practical Guide for Implementing Toyota's 4Ps*. McGraw-Hill, 2006.
8 Disney Imagineers. *The Imagineering Workout*. Disney Editions, 2005.

4 Need to Integrate Projects

Raúl Molteni

COORDINATION AND INTEGRATION OF PROJECTS AND METHODOLOGIES

We operate in a dynamic environment, where change is the norm. It is essential to create, develop, and adhere to projects that facilitate change. However, we also create work environments that are counterproductive to the effective functioning of the organization. Projects often compete with one another for priority, importance, budget, and attention at the operational level. Furthermore, projects are often pitted against day-to-day operations. Those responsible for implementing changes are under constant pressure; simultaneously, they also have to address the demands of day-to-day operations.

The answer, in general, is that "everything is important," which is an effective way to generate more confusion and conflict. While operational staff must prioritize and allocate resources to meet both requirements, management levels do not provide the necessary structure, guidance, or resources to support this.

Effective coordination and integration of projects with one another and with day-to-day operations requires co-creation—for the design—and collaboration—for the implementation— across functions and hierarchical levels.

Forrest Breyfogle[1] invites to integrate, improve, and align other initiatives such as Total Quality Management (TQM), ISO 9000, Malcolm Baldrige Assessments, and the Shingo Prize. This is critical because:

- They are indeed related and complementary.
- They are presented as distinct projects and programs. From the perspective of business professionals, managers, and employees within organizations, each of these components is perceived through different channels, with varying messages and priorities, and as a result, they are in competition with one another! So Design Thinking seems to be overtaking Agile, Agile seems to be overtaking Operational Excellence, the latter is overtaking Lean and Six Sigma, which were already competing with each other.

DOI: 10.1201/9781003544807-6

In order to "continuously improve confusion," there is also discussion of the necessity to develop leadership, project management, change management, and numerous other skills.

> *From the perspective of business professionals, managers, and employees within organizational structures, each of these "programs" is perceived to arrive through disparate channels, with disparate messages and priorities. Consequently, they are in competition with one another!*

A number of project management methodologies have developed frameworks, methodologies, and processes that facilitate the implementation of changes, and in a manner that is both effective and efficient. In the majority of cases, they pursue to minimize resistance and increase buy-in through the training of their practitioners and instances of their methodologies.

However, these actions do not cover all the necessary aspects.

It Is Not a Matter of Adding Actions to the Same Gantt

It is the understanding that decisions regarding each of the perspectives—technical and social— are mutually dependent. It is about integrating perspectives to:

- Consider the technical purpose and how it will be achieved.
- Determine how to manage the impact on people, including resistance, saturation, and conflicts.
- Manage the project itself, including planning, execution, and control.

CHANGE FROM THE "TECHNICAL" PERSPECTIVE

The project results in changes to the design, operation, control, transportation, and delivery of products and services. These changes are made using methodologies such as Six Sigma, Lean, Agile, Design Thinking, Customer Experience, Employee Experience, and ERP upgrade.

CHANGE FROM THE "SOCIAL" PERSPECTIVE

These projects have an impact on the individuals within the organization. These social consequences are frequently overlooked by those who are primarily focused on the technical aspects. The impact is due to the fact that they propose changes in the way people act and interact with one another in the workplace. **They alter the pattern in which individuals respond**. These technical changes propose a shift in the established "way things are done," suggesting a different approach. Such changes require us to step outside of our comfort zone, to move beyond our usual patterns of thought and action.

IS ALL OF THIS REALLY NECESSARY?

Is it essential to consider social aspects when implementing change? If the rules of how to implement Six Sigma Lean, Agile, or SAP are so clear, why add change management?

Nicholas William Leeson is a former trader who is most notable for causing the bankruptcy of Baring Bank, the oldest banking company in the UK, in 1995. During his visit to Buenos Aires a few years ago, he provided a detailed account of his actions. A question was posed to the group: "Given the numerous controls in place, how did this situation arise?" He responded, "Inadequate systems, ineffective controls, and the inappropriate placement of personnel. In essence, it is all about people." An environment of financial systems, procedures, and controls that are driven by people. New results and greater efficiency are not possible unless people change their behavior. Choosing technology first will not get you very far if no one wants to use that technology. It is also unrealistic to believe that you are really accomplishing with the requirements of a standard when you have a month to go before the audit and have to catch up on everything that has been neglected for months.

In addition, change is accelerating. The time to "adapt" to technology-induced changes is shrinking, creating enthusiasm for developers but concern and uncertainty for the rest. Getting back to our projects, times are also accelerating.

To improve a process, we use methodologies: Process Diagram, 5 Why, Hypothesis Testing, Poka-Yoke, 5S. To manage a project, we use methodologies such as PMI. To install SAP, we use predefined steps. But for a critical factor like social perspective, we use only training and information—usually called communication.

Change threatens us, no matter how close we are to those who promote it. Uncertainty creates insecurity, fear, associations, dissociations, and rejection of the new. In short, disorder. And disorder needs a framework and a way of working.

This is not about preparing for change management; it is about preparing for THE changes that are inevitably on the horizon.

NOTE

1   https://smartersolutions.com/about-smarter-solutions/forrest-breyfogle-iii/.

5 About Continuous Learning

Raúl Molteni and Carlos Lucena

FUNDAMENTAL PRINCIPLE: QUALITY

Prior to sharing key aspects of learning, it is essential to understand that the quality of learning is not solely determined by satisfaction surveys administered at the conclusion of training activities. The focus should be on the quality of the learning, not on the instructor. It is about the quality of the new possibilities that open up for the learner. What is crucial is whether it has enabled the trainee **to enhance their competencies and fulfill the role required by the project.**

> *Learning is about the quality of the new possibilities that open up for the learner.*

WHAT IS LEARNING?

It is the process of moving from a familiar situation to one that presents a challenge in the face of the unknown.[1] **It is an expansion of one's capacity to understand, relate to, and act in new ways**. Fredy Kofman defines learning as "the process of acquiring new skills that enable the achievement of previously unattainable objectives."

Learning requires leaving a comfort zone. To walk through a zone that shows us the weaknesses of today, that shows us the "mistakes" of the past, that attacks our ego by exposing us for having affirmed or acted in a way that learning shows us to be wrong or not the best today. Humility is a basic requirement for learning.

> *Humility is a basic requirement for learning; hence the importance of saying "I don't know" as an opening for learning.*

Here is a key to change projects that have a significant impact on people. Mistakes resulting from the learning process should be seen as a new instance of learning. If not, defensive behaviors will emerge, ignorance will be hidden, and the person will return to his or her comfort zone.

DOI: 10.1201/9781003544807-7

LEARNING BEGINS WITH KNOWING YOURSELF

The basis of learning lies in understanding what Heisenberg astonished the world with in 1927 when he formulated his uncertainty principle. And what others such as Humberto Maturana and Rafael Echeverría maintain:

> human beings cannot even know what is 'really real.' We only know how we see it, given our beliefs, experiences, and knowledge. Instead of striving to determine what is correct, we are left to admit the interpretations of the real world and seek those that are most useful for a particular purpose, knowing that there is no unambiguously correct interpretation.

The distinction is not between those who make mistakes and those who do not. The distinction is between those who learn from their mistakes and those who do not.

This understanding is what allows us to learn. It is admitting that there are other interpretations of what is happening and why it is happening. This openness is necessary. Learning begins with knowing yourself.

Learning begins with knowing yourself.

Learning requires that we accept that our explanations and interpretations of the quality and efficiency of what we do are our judgments against the potential judgments of others.

To learn, we must analyze together and accept the other person's interpretation of a situation. It is a judgment of the other person as well as our own.

We need intelligent conversations[2] and reflective practices, theories that explain whether our project plan achieved the goals we set and then validate and support them with facts and data. The stubbornness to defend what we believe to be the "truth" or "our" results and the lack of adequate practices to work as a team can lead us to erroneous conclusions. Thus, there are those who would call the project a failure and others who would call it a true success. Some would try to slow it down or reformulate it, and others would defend its continuity. What should be an intelligent task of analysis and learning becomes a conflict of the deaf and of power. An organization with people with an IQ of 140 forming an organization with people with an IQ of 85.

What should be an intelligent task of analysis and learning becomes a conflict of the deaf and of power. An organization with people with an IQ of 140 forming an organization with people with an IQ of 85.

STOP PRODUCING AND LEARN

You have to stop producing in order to keep producing at a good level. If you do not work on it, the capacity to produce is slowly depleted. On a day-to-day basis, we focus on producing, selling, assembling, reporting, and developing the project. It has the positive aspect—huge— of allowing the organization to exist.

But it carries a huge risk: we do not look into the future to see what we need to continue to be effective. We do not waste time trying to improve things that are "going well." And then one day we realize that we are out of date, that our people speak a language we do not understand, that the processes we have been working on are no longer sufficient to achieve what is needed, or that we urgently need a project for which we are not prepared.

READINESS TO LEARN

I already shared the 2-6-2. Now I share the other lesson. It was explained that in their journey to continuous improvement, the executives began to see themselves as getting worse and worse; they found defects and opportunities in quantity; however, customers began to see them as getting better and better. Until late in their improvement process, both they and their customers agreed that they were getting better and better. Why did they look worse and worse when they started to improve? Improvement involves self-criticism. The mistakes and opportunities were there, but they didn't see them.

This is one of the keys to learning. Looking for opportunities to improve, even when the results seem to show the opposite. **Learning in organizations means constantly testing experiences and transforming those experiences into knowledge** that is then accessible to the entire organization.

I like W. E. Deming's "subtle" shift to Walter Shewhart's PDCA Cycle. The Check is really about learning. It is about learning to extend knowledge to other processes, practices, and industries. Knowing that something worked does not ensure understanding of why it worked.

AN ENDEMIC PROBLEM

The endemic problem is trying to learn and improve by copying the "successful," going from fad to fad.

WHAT TO LEARN

Learning throughout a project involves three distinct focuses.

- *Skills.* Usually, to perform tasks that have been redesigned or are new.
- *Cognitive Process.* Related to the ability to analyze information and make decisions.
- *Social Process.* Relates to interaction, communication, and changes in behavior patterns that are part of the change.

HOW WE LEARN

Throughout my career, I have seen lasting and significant changes with different approaches. However, I agree and adhere to first creating change in "doing"—processes, procedures, and, as a consequence, routines. Learning by doing. Making mistakes, receiving feedback, stimulating discussion about the needs, difficulties, and benefits of applying knowledge.

We speak of individual learning when people learn and increase their ability to operate, interact, and behave more effectively and consistently. We speak of organizational learning when individual learning, through exchange and interaction, creates a more effective organization.

INDIVIDUAL LEARNING

It is important to recognize that we all have different learning styles. Our cognitive processes are influenced by our individual strengths and weaknesses. Understanding the way of learning that is closest to one's own favors accelerates the acquisition of new knowledge and skills.

LEARNING ORGANIZATION

We can speak of a learning organization when

- The learning is nourished by core values, by a sense of purpose, by external learning.
- Systems and processes are defined to generate learning.
- The structure contains elements that facilitate it, such as group work and shared experimentation that led to learning with others.
- Experiences—positive and not so positive—are studied and shared.

WATCHING AND ASKING

Gemba and Genchi Gembtuzu are two concepts that are repeatedly found in Lean. Go to the place and watch. Watching what happens and why it happens makes it possible to understand, for example, an operator's difficulty in applying a new procedure; or the gesture and emotion of a customer when talked about a product change or the response to his/her complaint. Watching and trying out leads to a deeper understanding and facilitates learning.

Visit the location in question to gain a better understanding of the situation. This will allow you to see what is happening and why it is happening.

LEARNING PROCESS

The learning process itself must be continuously improved based on rigorous analysis and the subsequent outcomes. As Jim Jaskol[3] states, "It's not uncommon for us to completely rethink our plans, reexamine our requirements, and go back to the drawing board in an attempt to have it all. Don't be afraid of saying 'it didn't work'; without that, there is no way we can learn."

> *Don't be afraid of saying "it didn't work."*

Michael Bungay Stanier[4], stands out as the essentials of the role of forgetting in learning. You can't learn new things, whether they are concepts ot tools, if you don't leave behind what was considered correct until then. And as with communication, teaching new behaviors and learning reflections requires persistence, repetition, and abundance.

> *Learning and communication are ongoing processes that require continuous attention. There is always room for improvement.*

THE CONCEPT OF FIRST- AND SECOND-ORDER LEARNING

It is important to distinguish between change and improvement. This also applies to first- and second-order learning.

> *If someone gets rich by winning the lottery, he has achieved something extraordinary, but he has not expanded his ability to win future lotteries.[5]*

Chris Argyris defined it as first and second loop learning. The first order of learning is based on the action taken in response to the results. The second order of learning is based on the knowledge, beliefs, and paradigms that led to the initial action.

To illustrate, consider the backlog for a sprint that has been planned but not yet completed. First-degree learning occurs when we simply change the way we plan the next sprint. Second-degree learning involves questioning the underlying assumptions, beliefs, and principles that support the analysis of the problem. It is a more profound process that delves into the tacit "truths" that we often take for granted. It is like the intention of the 5 Whys, which aims to identify the root cause of the issue at hand. However, the 5 Whys is typically employed when discussing process-related causes, not our mental processes.

> *The real learning is in going back and reflecting on the beliefs and paradigms that keep us from getting the results we want. Not just changing what we have done.*

First-degree learning involves modifying our existing processes to enhance performance. Second-degree learning entails reassessing the underlying assumptions and
beliefs that inform our decision making and equips us to navigate similar scenarios
with greater insight.

> *To learn to change, to improve our competencies to face change—not just the*
> *current change—we need deep learning, second-order learning.*

BARRIERS TO LEARNING

The barriers to learning are formidable. For example, feedback does not flow through
the filters of power, status, and authority.[6] "We have nothing to learn," "We do it better
than anyone else," or "It's 'somebody's name' fault" are common.

HERE ARE SOME BARRIERS THAT NEED TO BE ADDRESSED

- *Intending to do and learn everything.* Intending to learn and therefore act
  on a large number of issues or opportunities. Learn focusing on the great
  opportunity, about the great insight. This implies that something has to be de-
  prioritized and postponed.
- *Not saying "I don't know."* Avoiding admission of ignorance when we are not
  able to, or if the environment severely punishes not knowing.
- *Mistaking knowledge for absolute truth.* Learning entails recognizing gaps in
  one's knowledge. To know is also to recognize that there is no single answer
  and that there are different possibilities for action. Not guaranteeing something does not imply a lack of knowledge. In fact, it could signify a high level
  of understanding.
- *Failing to invest the necessary time, effort, and resources.* As W. E. Deming
  put it, "If you think training is expensive, try ignorance."
- *Not accepting authority from others.* This is a common barrier when it comes
  to learning at the management and leadership levels. It is important to accept
  that someone lower in the hierarchy can teach someone at the executive level,
  particularly when it comes to specific subject matter expertise.

NOTES

1 Generación Mas, *Guía de Estudio - Aprendizaje*. Generación Mas -www.generacionmas.
   com-, 2023]
2 Peter M. Senge. *The Fifth Discipline: The Art & Practice of The Learning Organization*.
   Deckle Edge. 2006.
3 Jim Jaskol, *The Imagineering Way, Ideas to Ignite Your Creativity*. Disney Editions, 2003.

4   Michael Bungay Stanier, *The Coaching Habit, Say Less, Ask More & Change the Way You Lead Forever*. Crayons Press, 2016.
5   Peter M. Senge, Art. Kleiner, Charlotte Roberts, Ricchard Ross, Bryan Smith, *The Fifth Discipline in Practice*. Doubleday, 1994.
6   Philip Sadler, Designing Organizations, *the Foundation for Excellence*. Kogan Page, 1998.

6 About Continuous Communication

Raúl Molteni and Carlos Lucena

FUNDAMENTAL PRINCIPLE: QUALITY

Before sharing key aspects of communication, it is essential to understand that, in a project, quality communication leads to more than just a better understanding between the parties. Quality communication enables a deeper exploration of diverse perspectives and facilitates the improvement of the initial exchange.

> *Quality communication fosters mutual understanding between parties, enabling them to elevate the issue to a more profound level.*

IT IS NOT SIMPLY INFORMATION; IT IS COMMUNICATION!

It is not simply a matter of conveying information; it is about effective communication.

Communication is not a "one-way" process; it is a "back-and-forth" exchange between the parties. It is not a unilateral act where one person emits a message and the other person simply understands it. Rather, it is a mutual process where both parties emit and both parties understand. This exchange requires shared responsibilities rather than just one party taking on the entire burden.

LISTENING TO COMMUNICATE

Effective communication requires a proactive, generous, and open-minded approach to listening, free from any preconceptions. Responding is not sufficient; one must reflect. It is important to note that communication should also occur where the opinions and concerns of the staff regarding the project and change are expressed. The response to these opinions and concerns can significantly impact the mood, confidence, expectations, and level of adhesion to the change.

> *Communication is not a one-way process; it is a back-and-forth exchange between the parties. It is not a unilateral act where one person emits a message*

 DOI: 10.1201/9781003544807-8

and the other person simply understands it. Rather, it is a mutual process where the parties emit and understand. This exchange requires shared responsibilities rather than just one party taking on the entire burden.

Communicating requires paying attention to one's own words, emotions, and gestures. Listening requires forgetting all of our own messages, emotions, and judgments, paying attention, and asking questions to understand the other person's intent.

Part of communication is monitoring the understanding of messages at all levels, not just the groups to whom they are addressed. A poorly understood message about the project's objective in one unit of the organization will be transmitted to the rest of the units, creating future difficulties in implementing the project.

Monitor the understanding of messages at all levels, not just in the groups to which they are directed.

For example, communicating the vision, plans, and responsibilities must include understanding and considering the recipients' reactions and perceptions.

Between what I think, what I want to say, what I think I say, what I say, what you want to hear, what you hear, what you think you understand, what you want to understand, and what you understand, there are nine ways of not understanding each other.[1]

This happens throughout the project. The project itself will create new concerns, perceptions, and needs, which in turn will create new communication needs and opportunities. So, communication is not something that has to be done at a certain stage. It is definitely something that must begin before the project becomes official and be maintained until after the project is completed.

It is not something to be done at a certain stage. It is definitely something that starts before the project is made official and must be maintained until after the project is completed.

PRINCIPLES OF GOOD COMMUNICATION

There is no magic formula for communication. The keys are as numerous as the elements involved: who is sending, what message they are sending, how they are sending it—through language, emotion, and physicality—through what medium they are sending it. And, as I said, the same elements play for the reception. The following five principles represent a checklist for effective communication:

- *Quantity*. Keep your messages brief and precise.

- *Quality*. Provide information that is accurate and verifiable. The communication should be sincere.
- *Relevance*. Give information related to the key aspects.
- *Modality*. Make the process straightforward. Give clear and unambiguous expressions.
- *Timeliness*. The project needs it at the moment.

> *Brief, sincere, simple, relevant, and timely.*

The most challenging aspect is active listening and reflecting on what is heard. This requires careful planning and preparation. Listening is an art form, as Bradlee Snow has noted.[2]

> *The most challenging aspect is actively listening and reflecting on what has been heard.*

LANGUAGE, CORPORALITY, AND EMOTIONS

Communication is not solely reliant on language. In addition to language, messages are conveyed through emotional expression and body language. A recognition or lack thereof, a dismissal or hiring, a promotion or nonpromotion, budget allocation or budget cut are also forms of communication. It is often the case that nonverbal communication is more persuasive than verbal or written communication.

> *Communication is not solely reliant on language. Furthermore, messages are conveyed through emotional cues and body language.*

THE IMPORTANCE OF WHO COMMUNICATES AND WHEN

WHO COMMUNICATES

Choosing who communicates is key. Who communicates, and how is it received and viewed by other employees? Is he seen as trustworthy? Are they simply relaying messages written by others? Are they sincere? Can they stand by what they say?

ON SINCERITY

Sincerity and transparency in communication are especially important. Much-used mechanisms such as a video or presentation by a CEO, director, or owner proclaiming the need for change have little life if it is an empty message without continuity.

ABOUT THE CONTEXT

When people are approached to gain their commitment, the focus is on how the change will affect them, regardless of its importance to the organization. If continuity is sought, the focus will shift to the relationship between personal efforts and the impact they may or may not have on the organization.

REGARDING THE MESSAGE

People's perceptions, experiences with change, and emotions should guide the definition of messages and interlocutors. The views of the change leaders themselves should not always be the focus.

People's perceptions, experiences with change, and emotions should guide the definition of messages and interlocutors.

PLANNED AND UNPLANNED COMMUNICATION

While it may appear to be a play on words, "planned" and "unplanned" communication need to be planned.

PLANNED COMMUNICATION

Planned communication is the process of defining the objectives, strategy, and key messages for engaging with stakeholders in a conversation. Aspects related to vision, roles, major milestones, training, pilot tests, results, inhibitors, and other benefits are to be communicated and can be defined in advance. Anticipation is necessary for this process.

EMERGING COMMUNICATION

The plan should also anticipate when unplanned communication will be needed, including what, who, and with whom to communicate in the face of an unforeseen event. A union announcement, an accident, a rumor, a poorly received communication, or a clear example of positive behavior for the project and change may require action. An unexpected event might trigger the need for unplanned communication.

A method must be developed to address communication needs in a prompt and effective manner, given the dynamic nature of change.

It is important to be aware that, even in the case of a small group, the contradictory "message" of what the change intends can reach the whole unit.

NOTES

1 Source unknown.
2 Disney Imagineers, *The Imagineering Workout*. Disney Editions, 2005.

7 About Coaching and Mentoring

Raúl Molteni and Carlos Lucena

COMMUNICATION AND TRAINING ARE NOT ENOUGH

It is insufficient to rely on communication and training alone to achieve new sustainable behaviors. To reinforce awareness, commitment, and the integration of new knowledge and competencies into habits and routines, additional tools, motivators, and incentives are necessary. To ensure continuity, a process must be established to address questions such as "What's in it for me?"

WHAT ELSE IS NEEDED

Change promoters and the rest of those affected need to be supported before, during, and after the change. Who accompanies them in these reflections? Who can help them ask questions that will help them see their own potential and overcome their fears? Who helps directors, managers, and middle managers to reflect on and understand the changes they need to make? How do you help those who promote change to understand that resistance is to be expected and not the result of people's bad intentions? How do you help them understand the perspective of those who have been working, and probably congratulated and promoted, in a way that seems not to be the expected one anymore? How do you help everyone understand one another and find a win-win? Who will help everyone go through the change with as little stress and conflict as possible and with as much mutual benefit as possible?

> *Change promoters and the rest of those affected need to be supported before, during, and after the project.*

This methodical support and assistance is what I call coaching and mentoring, especially for the stages of Commitment, Knowledge, and Continuity—see Chapter 10, "The Approach." A role that, through questions (coaching) and questions, reflections, and suggestions (mentoring) makes it possible to find new ways of acting in the bitter and difficult situations that projects and their changes bring about.

DOI: 10.1201/9781003544807-9

THE CONTRIBUTION OF COACHING

The European School of Coaching defines coaching as a contribution that includes "The art of asking questions to assist others in learning and exploring new beliefs that result in the achievement of their goals."[1] It is based on the ability to observe what someone is saying with the purpose of not only knowing what is being said, but also to know (interpret) the soul (understood as the particular way of being) of the speaker.[2] It is not just containment. It is accompaniment, so that people find more and better ways of acting in the face of a change that affects them and probably creates an emotional "tsunami."

> *It is not just containment. It is accompaniment, so that people find more and better ways of acting in the face of a change that affects them and probably creates an emotional "tsunami."*

FROM A CERTIFIED PROFESSIONAL COACH—LORENA MARÍA GÓMEZ[3]

Lorena es a *Bachelor's Degree in Social Communication and Professional Ontological Coach*. One of the most important aspects of coaching, according to Lorena, is to redefine the concept of learning. From the perspective of language ontology, "knowing" entails not only knowledge but also experience gained through repeated application. The objective is to develop the ability to apply knowledge effectively in a practical setting. Ontological Coaching views learning as the integration of a new "know-how" to act. Unfolding potential requires an introspective look at our existing capabilities. It involves identifying areas where we are not fully utilizing our resources, whether due to omission or lack of awareness.

Since we are "thrown into the world," we interact with the context and with other human beings, with the abilities that are inherent to our species: we observe, listen, perceive, feel, and almost instinctively interact with the environment in search of the desired results. When we achieve the desired outcome, we tend to repeat those actions and build our comfort zone. What happens when we want something else or when we do not get what we want?

We seek to learn new things. Through learning we create change and evolve. Ontological Coaching is a learning process that combines the dimensions of being and doing. We access the outside world from the "observers we are," filtering the information we receive based on our past experiences, family, culture, education, social practices, expectations, which act as the background of the "story we tell ourselves" and which turns out to be a valid view, but not the only one.

The Ontological Coach is an observer who facilitates these transformational learning spaces, interacting with the "being" from the language, thought/word, emotionality, and corporeality, with the objective of expanding the capacity to "do." As a facilitator, the Ontological Coach encourages reflection, guides us in transforming the reality we create for ourselves, challenges us to question limiting beliefs, and makes us investigate so that we can distinguish new perspectives that allow us to generate new interpretations.

By continuously learning, leaders and collaborators can leverage crises as opportunities for improvement, cultivate adaptability for rapid decision-making, and empower themselves by coordinating actions effectively to achieve superior results.

> *Our ability to create something that did not previously exist for us is the key to our achievements. It is about inventing new interpretations and practices to make our reality consistent with our commitments. Jim Selman*

OUR PERCEPTION OF THE PROJECT AND CHANGE IS INVOLUNTARY

In the same way that a situation, expression, smell, or image reminds us of something from our childhood and creates a positive or negative emotion, the exposure of the project and the resulting changes create involuntary positive and negative reactions. What can be voluntary is our response. Since we are unconscious, how can we make the reaction voluntary? With a mirror that reflects back to us what we say to ourselves and the emotion it generates, and that allows us to reflect and find the differences between what is happening and what we think is happening. This is what the coach helps us to do. The mentor adds his or her experience and expertise in the form of feedback.

> *The internal dialogue about a project and change generates emotions, moods, opinions, and reactions. They are unconscious.*

We think this change will hurt us like the previous one. Will it? We think this change will be just like the previous one. Will it? We think "improving a process by doing Lean" is for Toyota and "being Agile" is for Spotify. Wouldn't it work for us? "There is no benefit to me in committing to this change." Isn't there?

A lot of the conflicts that arise in change processes—standardization, improvement, design, transformation, digitalization, agility—are caused by misinterpreting what others say and do.

WHY COACHING AND MENTORING

It is challenging to reflect on one's self-perception. It requires guidance, support, an objective external perspective, and a different kind of observer. This is where coaches and mentors can play a valuable role.

> *It is challenging to reflect on one's self-perception. It requires guidance, support, an objective external perspective, and a different kind of observer.*

Both those who promote change and those who "suffer" from it need support to find a win-win situation. For some, this means designing and developing a project—and the resulting change—that maximizes benefits and minimizes conflict; for others, it means finding their place and helping them go through the change with less pain, stress, and complexity. The objective is to provide support and minimize:

- Resistance: to encourage as many individuals as possible to adopt the required behaviors as a habit in the shortest possible time and in an integrated manner (organizational vision).
- The impact of stress, demotivation, frustration, helplessness, and complexity that changes may generate in the personnel at all levels (individual vision).

REGARDING MANAGERS AND SPONSORS

Providing this support to everyone affected by the change is impossible. However, there are roles that cannot be left out: the management and sponsorship levels.

Middle managers have a very strong and specific role to play in a time of change. On the one hand, they must be considered as the other employees of the organization, working with them to understand the what for, the why of change, the why now, and how the change will affect their work. They need to be ready, motivated, and supportive of the change. On the other hand, they must be informed and trained to work with their employees on the change. They are key communicators, should receive frontline feedback, and must answer their employees' questions about the change. This dual role and its demands generate stronger and more complex resistance than other groups.

> *Coach them to understand what for, why the change, why now, and how the change will affect their work. On the other hand, they need to be informed and trained to work with their employees on the change.*

In the case of sponsors, who are typically managers or directors, the impression is that their role is to answer questions and make decisions. The sponsor role is more about guiding, asking, and helping others to reflect, identify inhibitors, and remove them. The Coaching Plan should help reflect on current behaviors and find satisfaction and motivation with new behaviors.

> *The help is focused on each person's specific needs, their personal scenario, role, fears, and dilemmas.*

THE ROLE OF THE BOSS AND THE COACHING KATA

I believe that the Coaching Kata is an excellent model: "The primary task of Toyota's managers and leaders does not revolve around improvement per se, but around increasing the improvement capability of people."[4] It is about ensuring that what people understand, learn, and the way they think are validated as they apply it on a daily basis.

> *"The primary task of Toyota's managers and leaders does not revolve around improvement per se, but around increasing the improvement capability of people.*

The Coaching Kata is a structured dialogue between a mentor and a mentee, comprising key elements:

- Let the mentee—coachee—discover things for himself. The mentor asks the mentee questions, not to lead him to a certain solution—even if he thinks so—but to understand how he thinks, how he approaches the situation, and how he finds the solution himself.
- The goal is for the mentor—coach—to lead the mentee—coachee—to discover and learn by himself and through his mistakes.
- The mentee is responsible for the action, and the mentor shares responsibility for the results.

> *Let the mentee—coachee—discover things for himself and learn by doing.*

In the Coaching Kata, through dialogue and conversation, the mentor gains insight into the mentee's thought process, planning methodology, and analytical approach. In addition, the mentor learns about the steps the mentee is taking and the areas where the mentee requires support. The chapter "Continuous Support" discusses the integration of this support into the overall project plan.

NOTES

1   European School of Coaching.
2   Rafael Echeverría, *Ontology of Language*. Comunicaciones Noreste, 2003.
3   Lorena María Gómez. *Bachelor's Degree in Social Communication and Professional Ontological Coach.*
4   Mike Rother, *Toyota Kata, Managing People for Improvement, Adaptiveness, and Superior Results*. McGraw-Hill, 2010.

8 About Owners, the Board, and CEOs

Raúl Molteni

OWNERS, BOARD, AND EXECUTIVES AND THEIR CRITICAL ROLE

It is generally understood that the board of directors plays a critical role in overseeing management and ensuring that the company operates in full compliance with all legal and regulatory requirements. It is also understood, though not always realized, that the board is responsible for setting the strategic direction of the organization, overseeing risk management, setting and ensuring that management sets clear expectations for ethical behavior and transparency, establishing a code of conduct and clear policies and standards, among other things.

The board should step forward and realize its contribution as much more than ensuring compliance or overseeing management. It's about ensuring the sustainability of the organization. This includes (1) purpose, in the spirit of service to stakeholders and society at large, and the practice of "corporate" citizenship; (2) strategic direction, including the vision for the future, key goals and strategies, and the overall improvement and innovation system; and (3) culture, encompassing the values, beliefs, and principles that underlie a collective way of thinking, feeling, and judging that reinforces commitment and is critical in shaping decisions and actions.[1]

STRATEGIC, INTEGRAL, AND EVOLUTIONARY VISION

Strategic planning without strategic thinking is useless. The board should be judged based on the organization's ability and potential to sustain and advance the business. Their responsibility should go beyond approving or questioning management's plans and ensuring compliance with regulations and laws—not because that is not necessary and important. It is critical that they and the organization learn to change.

> *It is critical that the board and the organization learn to change, not just wait for others to change.*

DOI: 10.1201/9781003544807-10

Suppose that you are planning a change that will impact your business. In this case, you will probably need to change mindsets and technologies, find new ways of conceiving, thinking about, and using data, new ways of looking at and considering other people, change the criteria used to make decisions, find a new way of thinking about a problem or how to improve your results.

You'll also find that most people won't follow an idea just because it's logical, so how they feel about it will be a key issue. No matter what change you are talking about, you will most likely need to consider how it affects people, their minds and hearts, and the culture of the organization. This is not an HR issue. This is a leadership issue. You can't replace your conviction and determination for the project.

No single function can create the strategic and business-level sense of urgency that a CEO, owner, or board member can. Or create a motivational vision, enlist volunteers, remove potential inhibitors and barriers, and sustain the change as best they can.

> *"If you don't believe it, sooner or later one of your decisions will show what you really believe." Anonymous.*

Frederick Taylor recommended that the best way to manage manufacturing was to standardize the activities of workers into simple, repetitive tasks and then closely supervise them.[2] Operators were "doers," and managers were "planners." Productivity gains were significant in the first half of the 20th century. Over time, the need for greater interaction with employees arose, and the era of talk about participation, autonomy, and teamwork began.

Some have criticized Taylor's guidelines. A practice that was highly effective in a different context is now being questioned and criticized. It is being replaced. There is no evidence of evolution or improvement; rather, there is a clear indication of obsolescence. "Old" solutions are questioned and criticized. This is precisely what I observe in organizations. The necessity of change and the appropriate times for transition are called into question. Fads are embraced. Design Thinking diverts attention from Agile, Agile from Lean, Lean from Six Sigma, and so on. Management replaces focus by introducing discontinuity instead of facilitating evolution. A new CEO or unit manager must establish their own imprint, program, and success. It would be beneficial if they built on existing structures rather than seeking to replace them. This creates a sense that, in time, this new approach will also be "thrown away" and replaced, and not improved or evolved.

ORGANIZATIONAL CULTURE

Organizational culture is much more critical than organizational structure, and a board is ultimately responsible for shaping that culture.[3] You can't build and install continuous improvement, agility, or transformation. You can install machines and build a plant, but not a culture. That requires an understanding and commitment that goes beyond asking someone to place an order or oversee construction.

OWNER, BOARD, AND EXECUTIVES AND CONTINUOUS IMPROVEMENT

When a project has an impact on the culture of the organization, it is natural for people to consider how long-lasting and how supported the project will be. "Will the directors approve this project?" What is the owner's position on this matter? "Will the project suffer the same fate as the one initiated under the previous CEO, which was then left unfinished upon his departure from the company?"

By taking action or not taking action, the board has a positive or negative impact on what management can do and accomplish. Especially in a family business. That's why it's necessary to have Quality at the Top if an organization wants to have sustainable quality projects, processes, products, services, experiences, and results.[4]

In 1968, Joseph Juran, one of the world's top-quality gurus, stated in his book *The Corporate Director*: "In the absence of a sincere manifestation of interest at the top, little will happen at the bottom ... ."

More than 50 years later, Philip Armstrong,[5] former head of the Global Corporate Governance Forum, International Finance Corporation, said in accepting the IAQ Marcos Bertin Quality in Governance Medal:

> Quality is an essential part of corporate governance. It's not just about making sure you have a nice report that shows you've complied with a whole bunch of standards that are either in some form of regulation or code or law, it's really about how the board really views corporate governance as a value-added principle of its business proposition.

The speed of technological, environmental, and social evolution and its impact on the way all types of organizations operate requires a review of the principles and practices that boards apply. Questions arise: Are your current board practices still effective and efficient enough? Should any be changed, discontinued, or added?

The answer is: A board should understand and encourage evolution and continuous improvement. It's not just a C-suite responsibility.

OWNER, BOARD, AND EXECUTIVES AND THEIR OWN LEARNING

Second-order learning[6] should be a fundamental weapon for the board and executives of an organization. Is the performance of the C-suite evaluated by analyzing the knowledge, beliefs, and values that led them to define those plans and make those decisions?

To tackle meaningful projects, especially those involving "changing the culture," second-order learning must start with the owner, board, CEO, or unit manager. The following questions should be answered:

- What beliefs, knowledge, paradigms, and experiences have led us to adopt previous projects that had produced the current results?
- What beliefs, knowledge, paradigms, and experiences lead us to adopt this project? Should they be the same? If not, why not?

THE SCOPE OF MANAGEMENT CONTROL

Joseph Juran, supported by research in different countries, has stated that 80% of errors made in an organization are controllable by management. The remaining 20% is controllable only by the operator of the process in which they are made. W. E. Deming states that 94% of the variation in system performance is caused by the system itself. In addition, he mentions a letter from his friend, Dr. Noriaki Kano,[7] who asserts that "the issues within the organization are not solely due to operator or supervisor error. It is unlikely that these operators are intentionally making mistakes. They are constrained by the system."

What is the system? Management style, values, beliefs, policies, strategy or lack thereof, investment and resource decisions, poor management, and staff training; processes, practices, and tasks that create a negative emotional response, including discomfort, boredom, and urgency; unfriendly computer systems that encourage the use of Excel spreadsheets. The reduction in maintenance costs leading to machines becoming unreliable.

One might concur with Dr. Kano's assessment. It could also be argued that people are sometimes indifferent to these issues. However, why should they be when their superiors are not?

Best efforts are not enough.[8]

WHAT AN EXECUTIVE SAYS IS NOT ENOUGH

The president and CEO of a medium-sized company has assumed control over one of the business units of an entrepreneurial group. The unit was performing well financially, with a higher profit margin than the industry average. It supplied companies in both the automotive and nonautomotive sectors, as well as exporting its products. A labor dispute between two factions of the union resulted in a divided workforce and prompted legal action against the company. The management team, which had negotiated highly favorable benefits for the personnel, was left feeling defrauded and at high risk as a result of the lawsuit. The business group shifted its investment priorities to other plants, and this one was no longer a strategic focus. Over time, preventive maintenance was phased out. Equipment became obsolete, productivity declined, and personnel training was eliminated.

Plant, maintenance, and production managers took over from one another. The reasons for the deterioration were never clear, and never convinced the CEO that it was not their fault. For example, one of the two critical pieces of equipment and the bottleneck of the plant was significantly reducing its productivity levels. The CEO decided to sell it and avoid the discussion about the real cause since none of the presented convinced him. Months later, a new plant manager proposed an improvement plan, but the administrative director—completely removed from the plant—did not like it, and the CEO backed the director. In a last-ditch effort to rally the entire staff behind an improvement campaign, the management team met over a weekend to develop a consensus turnaround plan. The plan generated enthusiasm and, within a month, showed improvement in the productivity of the critical equipment. At one of the weekly follow-up meetings, the CEO was present expressing great distrust in the individual and group's ability to achieve recovery.

A good executive, a good sponsor, and a good facilitator puts "what can I do" before "the fault lies with...".

His messages never failed to include phrases such as continuous improvement, the need to avoid mistakes, or participation. His verbal message was consistent with what he wanted. However, he communicated not only with words; tone of voice, gestures, emotions, and choices said much more, and in this case blocked any further actions of improvement.

It's not just the words that are used. It is also what a top executive communicates with his words, voice, gestures, and emotions.

THE NEED FOR EMPATHY

Instead of thinking about how to get people to change, start thinking about people.

Have you ever felt forced to accept a change you didn't believe in? Haven't your parents or other siblings ever been in this situation? A friend? Your wife or children? How did you or they feel? Always well and ready to accept it without concern? People who do not like change are not necessarily wrong, outdated, or bad. Our experiences, interests, and knowledge make us different. We all see the change from our perspective based on our beliefs, values, experiences, and knowledge.

Haven't you ever suggested a change to someone you love? Were you always wrong? Didn't they eventually discover that you were right, and that change was a good thing? Didn't they benefit in some way from the change? People who suggest change are not necessarily crazy, wrong, or trying to hurt you. Neither are people who resist change.

The owner, board, and C-suite of an organization that wants to implement a project that aims to evolve the way it does business and the results it achieves should use empathy when making decisions.

NOTES

1 International Academy for Quality, *Quality in Governance Think Tank, Quality at the Top, a Quality Guide for Boards.* 2023.
2 Frederick W. Taylor, *The Principles of Scientific Management,* Dover Publications. 1997.
3 International Academy for Quality, *Quality In Governance Think Tank, Quality at the Top, A Quality Guide for Boards.*
4 Raúl Molteni, President-Elect, International Academy for Quality, - *Quality in Governance Think Tank, Quality at the Top, A Quality Guide for Boards.*
5 Philip Armstrong, Former Head, Global Corporate Governance Forum, International Finance Corporation. (www.youtube.com/watch?v=z6FHerayPQE).
6 See Chapters "About Continuous Learning" and "Implementing and Learning".
7 He also mentions it in his book *Out of the Crisis*, Massachusetts Institute of Technology, 1986; see also Joseph M. Juran and Blanton A. Godfrey, *Juran's Quality Handbook.* McGraw-Hill, 1999; William E. Deming, *Out of the Crisis.* Massachusetts Institute of Technology, 1986.
8 W. E. Deming, *Out of the Crisis*, Massachusetts Institute of Technology, 1986.

9 The Role of HR

Raúl Molteni and Carlos Lucena

THE ROLE OF HR

Human Capital, HR, or Talent intervention is key to an effective approach to change. This could involve direct intervention in the teams that develop the projects or support for them.

In many organizations, the focus of HR—Human Resources or Talent, depending on what they call it—leaves an almost empty space in change management and the real employee experience. In our view, this focus deprives these areas of the strategic positioning they need.

SEGMENTING YOUR STAKEHOLDERS

In Chapter 10, "The Approach," I use the ADKAR Model[1] to describe the phases we go through to sustain a change.

1. Accepting the need, nature, and timing of the change.
2. Accepting that we are part of it and making a positive contribution.
3. Being able to implement the necessary practices and routines.
4. Having the ability and capacity to apply change consistently and sustainably.
5. Receiving and perceiving additional support from the organization.

It does not mean that we are positioned in one phase or the other, but that we evolve as if we have one foot in the previous element. It is not a one-way street, but a two-way street. The absence of organizational leadership causes a person to lose continuity, hesitate to commit, and ultimately question change.

Whoever they are, understand them. One by one. Segment and define the position, understanding, attitude, aptitude, and emotion of at least those people who have significant influence over others. The organization's attention and effort to have this segmented knowledge are a key factor in the success of the change.

 DOI: 10.1201/9781003544807-11

Knowing who is being conditioned and by whom is a key factor in identifying, preventing, and managing resistance.

BEING RIGOROUS

"Take your time, understand the causes, inhibitors, and opportunities." This message seems clear to Lean practitioners, Green Belts, Black Belts, PMs, and Scrum Masters.

However, the rigor is lost when dealing with social issues. For example, teams tend to be less rigorous in finding the root causes of change resistance and risks associated with social issues than they are in addressing processes, products, services, and plans. There is often a lack of understanding of the origin of resistance, which can lead to vague plans, training everyone equally, and providing information without precision. This can also result in confusing information and asking the sponsor to act without a clear understanding of what is required.

SUMMARY

The following is a summary of our suggestions for HR or equivalent functions, especially in small and medium-sized companies:

- *Allow and seek greater participation* in—or at least a greater understanding of—strategic decisions. Have a long-term vision.
- *Forget fads.* For example, installing applications and mechanisms that seem to be successful for other organizations, without focusing on real solutions to the problems that affect organizational health and culture. Work on creating the individual and organizational competencies your organization needs, not those that "successful companies" proclaim. Understand truly what "talent" means to the organization's present and future and make concrete plans to develop it in place.
- *Loosen the emphasis on theories and follow the "walk the talk."* Understand organizational health and provide segmented responses based on employee experience, not gender, age, or role segmentation.
- *Reinforce customer focus* as part of "walking the talk"—personally and organizationally. Assigning HR leaders to every other function does not ensure understanding and, much less, satisfaction of these functions with HR.
- Increase the *facilitation* and *change skills* of the people in the function.
- Help *integrate and coordinate projects and programs*, especially those of the function itself.
- *Forget the abstraction associated with "change."* Apply rigorous change management methodologies based on business needs. Prepare to manage change—current and future—not to manage a particular change.
- Collaborate with the installation of *ethical concerns* based on the organization's impact on society.

- *Use indicators* that make it possible to perceive changes in the mentality, behavior, and habits of personnel.

"Deciding what not to do is as important as deciding what to do." **Steve Jobs.**

NOTE

1   ® ADKAR and/or Prosci is a registered trademark of Prosci, Inc. All rights reserved.

Part III

The Path to a Robust Project

This section outlines the methodology and process for developing a project with a tangible impact on business outcomes and organizational culture.

DOI: 10.1201/9781003544807-12

10 The Approach

Raúl Molteni

ABOUT AN APPROACH

The Approach in our context refers to **a way of reading, understanding, and designing organic solutions, and more importantly, a guide** that sets out the basic concepts on which is built the method to the project management process with an integral view.

APPROACHES TO CHANGE

The relationship between beliefs, attitudes, and behavior has been, and continues to be, the subject of much research. One thing we take for granted is that what people say they would do in a given situation is not necessarily how they would behave. Let's take the example of quality and price. When asked, people may say that they would pay more for a higher quality product, but this is not always the case, at least in some countries. People may answer what they would like to do or what is politically correct, but you can't take that for granted.

The most common "approach" to change is based on doing nothing specific—ignoring it and then blaming people for not buying it—or relying on what can be done with information and training. Very different from this is the "Lean Approach" or Toyota Approach, which is clearly expressed by Jeffrey Liker and David Meier in *The Toyota Way Fieldbook*: "If we can change behavior, we can change attitudes ... the bottom line is that we're more likely to change what people think by changing what they do than to change what people do by changing what they think."[1] And he continues: "Direct experience, with immediate on-the-spot coaching and feedback, will change behavior over time. On the other hand, trying to change what people believe through persuasive speeches, interactive video courses, or training will not work." In the same vein, Kotter says, "Culture changes only after you have successfully changed people's actions, after the new behavior has produced some group benefit for a period of time."[2]

A complementary approach is to consider change from a personal perspective. We have self-talk and emotions about an event based on our beliefs, experiences, learning, and messages received during our lives. These conversations and emotions influence our interpretation of events and, subsequently, our reactions to them. This

DOI: 10.1201/9781003544807-13

approach aims to prompt reflection on these internal conversations, enabling individuals to recognize their influence on judgments and reactions.

TECHNICAL FRAMEWORKS

There are some similarities in the way some of them work.[3]

- Gain clarity about what is intended—whether it is to fix a problem, design a process, product or service, exploit an opportunity, or develop a project. Clear vision. Visible and inspiring leaders. Policies and practices for a sustainable environment.
- Involve people and care for employee engagement, experience, and development.
- Understand stakeholders' needs. Design, purchase, produce, and delivery for the best employee and customer experience. Efficiency for all shareholders' results and well-being.
- Analyze the baseline using data. Leverage data and use metrics to track results.
- Identify the reasons for the effects to be neutralized or the key aspects of what is to be achieved.
- Generate creative solutions. Technology at the service of operations and stakeholders.
- Implement in a controlled manner.
- Validate the effectiveness of the solution and its sustainability over time with experience and data.
- Lean, Agile, and Six Sigma, in particular, offer continuity with a new opportunity.

WHAT I SEEK TO ACHIEVE

Establish a model and process for all professionals who are involved in the management of standardization, improvement, development, design, digitalization, evolution, agility, and transformation projects, and for those who wish to approach and methodologically integrate the social aspect of change. My goal is for them to gain knowledge of the key concepts and clarity in terms of "What should I do" to get people's buy-in when working on the project.

Here is a model with a rationale these professionals find in their technical methodologies:

- Clear definition of problems and opportunities and achievable goals.
- Identification of the causes of problems using data.
- Generating solutions that selectively impact causes and opportunities while considering risks.
- Applying the concept of value to critically review the results achieved, understanding that a change is not necessarily an improvement.

The key is learning to manage change, not learning to manage "this" change. Growth is more production, more customers, and more turnover. Development is more like Russell Ackoff's postulate: "increase the capacity of a system to satisfy its own needs and the needs of others." Development is gaining the ability to do more as well as to do it differently. For example, it is producing more and different products or services, selling more, but in different markets and through different channels.

> *My proposal is not to gain the ability to manage a change, but to manage current and future changes.*

THE APPROACH

The approach is grounded in theory and experience. My perspective on integrated processes, products, and services management—design, standardization, and improvement—is based on what I have learned from Blanton Godfrey, Greg Watson, Liz Keim, Janak Mehta, Forrest W. Breyfogle, Lennart Sandholm, Lars Sörqvist, Beth Cudney, Anders Fundin, Jeffrey Liker, Peter Pande, and Mike Rother, among others. On the other hand, Edward Baker, John Kotter, Prosci, ChangeFirst, John Hayes, Jeffrey Hiatt, and Vincente Gonçalves have allowed me to learn aspects of change management. Others, such as Marcos Bertin[4] and Hugo Strachan, have enabled me to deepen aspects related to the board of an organization, the DEC Association[5] with the experience of customers and employees, and Kamal Munir[6] and Shahzad Ansari[7] have ensured that I understand better the concepts related to strategic thinking.

> *The greatest learning and differentiation come from integrating concepts.*

As with robots, the greatest learning and differentiation come from integrating concepts. Their evolution does not come from a substantial improvement in one area of technology. It comes from integrating advances in multiple technologies— artificial intelligence, mechatronics, navigation, sensors, object recognition, and information processing, among others. It comes from looking at the ecosystem and learning from it.

What I share in this book comes from the search for similar integration. It forms a structure that comes from my own experience with organizations operating in Argentina, Bolivia, Brazil, Chile, Ecuador, Germany, Paraguay, Peru, Spain, the United States, and Uruguay. I have experienced great satisfaction by receiving the consecutive gold, silver, and bronze awards in the International Team Excellence Award of the American Society for Quality for 10 years, which were obtained by project teams that were trained and coached by us. I have worked with service and manufacturing teams in large and mid-size companies, as well as in transformation processes in manufacturing and service companies.

I have seen projects that were far from achieving the desired results. I may have learned as much from them as from the successful ones, even though there were far fewer of them. All of them, to a greater or lesser extent, have contributed to the integrated suggestions I are offering here.

MODEL OF REFERENCE

From an operation point of view, I have found the ADKAR Model developed by Prosci to be most effective.

> *To help the organization to get more people, in less time, to adopt as many new practices and behaviors as possible, and to maintain the necessary routines with the changes proposed by the project.*

AN APPROACH WITH A FRAMEWORK FOR CHANGE BASED ON THE TECHNICAL FRAMEWORK

> *This same path allows us to address social aspects.*

REGARDING THE OBJECTIVE AND ITS INTENDED OUTCOME

Just as you need to understand the technical indicators or aspects of the business that require design or improvement, from a social perspective, you need to specify the desired change. This entails identifying specific elements of the vision that influence how people perceive, think, and act. It encompasses the experiences, emotions, and behaviors that should be maintained, altered, or introduced.

ABOUT THE INITIAL SITUATION

The characteristics of the culture will help us identify which aspects could facilitate the implementation of changes and which are likely to pose significant challenges. In addition, the level of preparation, such as the availability of competencies to apply methodologies or to sponsor a project, will indicate structural issues that require initial reinforcement.

In the context of improvement projects, it is essential to gain a deep understanding and segmentation of all stakeholders, as well as problems. The change will not impact everyone equally; as a result, it will be necessary to estimate the impact that the change has and will have on each of the subgroups.

REASONS FOR UNDESIRED OR DESIRED EFFECTS

Once aspects of culture that hinder or facilitate change have been identified, it is beneficial to consider how they influence the process of securing buy-in for a change initiative. What factors will prevent or encourage individuals to become aware of and then commit to the change? What factors will impact their ability to gain competence and behave as expected with regard to the change? What factors will facilitate the maintenance of these behaviors over time?

I am discussing both individual and organizational resistance. Which organizational policies will impede individuals from achieving Awareness, Commitment, Competency, and the continuity of new behaviors and attitudes? Which factors should be maintained and which should be modified or introduced?

On the Creation of Solutions

In the same way that solutions are designed to address the root causes or specifically target an element or activity of the process, product, or service, resources must be designed to overcome the obstacles that hinder the acquisition of Awareness, Commitment, Competencies, Continuity, and Organizational Leadership.

On the Implementation of Solutions

This stage involves the development of a strategy, an overall plan, and partial plans that must be derived from and integrated into the overall plan. The use of indicators is a crucial aspect of this stage. For instance, indicators that demonstrate the evolution of the degree of buy-in and adherence to new behaviors are invaluable.

On Validating Solutions

Other indicators should be used to understand the level of Awareness, Commitment, and Continuity of the intended change.

New Improvement Path

Planning, strategy, and continuous feedback are also required to maintain people's motivation, skills, experience, and commitment to successive changes.

AN APPROACH FOCUSED ON HOW PEOPLE CHANGE

My proposal for the change management process is based on the ADKAR[8] Model. I have designed indicators and structured the change management process around this model.

Awareness of the Need for Change

A basic requirement for the "first step" toward change is acceptance of the change itself, acceptance of the nature, the need, and the timing of the change for the organization. Usually, what makes people aware of the need for change has to do with who articulates the need for change, the clarity with which they do so, and the reasons they justify it.

Particularly for senior management, how the change aligns with the organization's vision and strategy, what the risks are, what the costs of not changing are, and what the specific objectives are.

> *The first goal for those driving and leading change: to position change as necessary, beneficial, and timely.*

INDIVIDUAL COMMITMENT

Commitment refers to the desire that someone must actively participate, agree to be part of the change, and contribute positively to the change. One might agree that the change is appropriate for the organization—be aware of the need—but for various reasons, not want or agree to be part of it.

The need to be more customer focused and understand the real needs of the customer may lead to agreement, but it may be thought that it is the other people who need to change. The change may suggest more participative and agile leadership, but some middle managers and executives may see themselves as examples of participation without being so. People nearing retirement may see a change as appropriate for the organization and yet decide to stay on the sidelines, not making an effort and just waiting to retire.

> *Second purpose for those who promote and lead change: to get people to commit and add their own actions to the organizational effort.*

KNOWLEDGE AND SKILLS

This refers to the knowledge, training, competencies, and skills someone has to fulfill their role in the change. It requires that they understand what to do, what to do differently, and what to stop doing. And that they have the skills to do it.

Someone can recognize the need for change, and even more, can promote it. He or she can also have the desire and the total predisposition to contribute. But his or her profile, personality, skills, and natural conditions may not allow him or her to perform adequately in the face of the intended change. Can someone who has always received the message that customers are useless complainers change to a truly customer-oriented behavior? Can someone who is convinced that he is the best at a task accept the opinion of all those who are part of a work team?

> *Third purpose for those promoting and leading a change: the competencies—knowledge and skills—to fulfill what the project requires or will require of them.*

Continuity and Routines

Despite demonstrating their capacity to drive change based on their dedication, expertise, and abilities, they may face challenges in maintaining new behaviors over time. Some of these challenges are self-imposed. A golfer may achieve a par on a hole, but this does not necessarily guarantee that they will be able to repeat this performance on subsequent attempts. Some other obstacles are introduced by the organization itself or by a supervisor. For example, you may be asked to work in a team, to join one, to attend and contribute to team meetings to improve a process, and, upon returning to your usual place of work, you may find your boss asking, "Will you get to work now?"

Willingness and ability do not necessarily guarantee continuity. This requires ongoing evaluation, effort, and maintenance. In some cases, it may be more important to focus on obstacles to motivation than trying to provide motivation itself.

> *Fourth purpose for those who facilitate and lead change: to support people's actions by removing the barriers they face and will face.*

Leadership and Organizational Commitment

Organizational commitment may seem assured because owners, directors, and managers have proposed the change. This is not enough. Organizational commitment requires additional and ongoing efforts to address organizational resistance.

The availability of resources for people to fulfill their roles, the alignment of policies and processes, the avoidance of conflicts between projects, not losing the momentum of change, and not wearing people down with too many unrelated projects are litmus tests for individual commitment to be sustained.

If this does not happen, those who have supported and committed to the change, developed their own knowledge and skills, and are fulfilling their role may perceive the change as a fad and mutate into a position of disengagement and frustration.

> *Fifth purpose for change facilitators and leaders: managing organizational resistance—see Chapter 25, "Managing Resistances."*

DIFFERENT METHODOLOGIES FRAMED AND INTEGRATED INTO ONE FULL SYSTEM

Methodologies and processes impact motivation, beliefs, and attitudes, often leading to resistance when implementing change. Similarly, motivation, beliefs, and attitudes significantly influence the way methodologies are used and their long-term results.

WHAT DOES INTEGRATION MEAN?

In many organizations, the technical aspects are typically managed by operational areas, while the social aspects are often handled by the HR. This results in the execution of two unconnected projects, which require the understanding and execution of the same areas and people.

Integrating the social and technical perspectives is not simply a matter of combining the actions of each perspective on the same Gantt chart. This implies that decisions regarding social and technical aspects are interdependent.

While I have discussed the parallels between the technical and social aspects of project development, it is important to note that the stages do not necessarily occur concurrently. The characteristics, ambition, objectives, development, and methodology of the project—such as Lean, Six Sigma, and Agile—serve as a trigger to identify potential reactions to change, including individual resistance and structural resistance. Such reactions indicate the need for adjustments to the technical perspective.

> *Integrating the social and technical perspectives is not a matter of having only one Gantt chart. This implies that the choice of how to proceed with regard to social and technical aspects is interdependent.*

We should strive to comprehend the dynamics of each type of change and work to refine the eventual sequencing and simultaneity.

For example, when implementing SAP updates, it may be more appropriate to begin training final users when the impact on the stakeholder culture and environment is deemed suitable, rather than waiting until integration tests are complete. Based on my experience, this is typically well before the timeframe typically planned by SAP consultants. Matrices used to select Six Sigma projects typically consider the potential readiness and willingness of sponsors, champions, and process leaders as criteria. However, a methodological assessment, rather than simply gathering the team's opinion, would provide more comprehensive information to make the decision.

My experience indicates that it is beneficial to begin the analysis of the change process prior to the initiation of the technical project planning. The key phase is the one when people perceive the change begins, and that is not when the kickoff is done but when people start listening to something about a change.

> *The experience of the people involved in the change begins when they receive information about it, not when the project kickoff is completed.*

AN APPROACH NOT ONLY ABOUT TRAINING AND COMMUNICATION

The reasons why people do not join a change or hide it must be addressed with other mechanisms than those of communication and training. If the resistance is due to a lack of resources to comply with the change—tools, computer systems, and time, among other things— or if the change is intended only by a manager and not by the entire management, or the change expects of a person something that does not coincide with his or her interests, then additional and different support measures are required.

Mentoring and coaching—well understood and done by professionals—can provide additional support for reflection and reinterpretation. Building a coalition of sponsors is much more powerful than appointing one sponsor for the project. Managing resistance—systematically and methodically anticipating and preventing it—is much more powerful than reacting to it.

Even more powerful is the fact that management is reviewing and revising policies and processes that support concepts inconsistent with those proposed by the change. Reallocating budgets and resources to support the change and redesigning, improving, or eliminating processes, procedures, and controls can powerfully motivate people to accept change and adopt new behaviors.

> *Systematic and methodical management and anticipation and prevention of resistance is much more powerful than reacting to it.*

PEOPLE MATTER, NOT JUST THE ORGANIZATION MATTERS

The objective of the project is to provide the optimal experience for all stakeholders. It is not sufficient to expect people to change, "buy-in," and adopt new behaviors as quickly as possible. This is the organization's perspective and represents a key objective.

The other perspective is that of individuals who were incorporated or promoted for reasons or for actions that are now deemed inappropriate. The organization is responsible for establishing policies, processes, and systems, and it is this same organization that is now implementing changes to them. While there is a need for change, it is not acceptable to imply that people "have to change" and blame them for the difficulties that may arise. People have the right to expect change management to facilitate their transition to the vision, helping them to minimize the pains and inconveniences of change.

> *We may agree with the need for change, but we do not agree with the sentiment that people "have to change," as if they are to blame. People, we believe, have a right to be truly considered by change management.*

Identifying people's resistance to develop actions that facilitate their compliance should have a twofold purpose: to help them understand the need and commit to act differently, and to ensure that this different action is achieved with a positive experience.

Communicating the need and opportunity for change should show the benefits to the organization as well as to the people. The phrase "what's in it for me" is a requirement for change management, not a sales pitch.

> *The expression "what's in it for me" is a requirement for change management, not a sales pitch.*

Monitoring people's experience of change is a responsibility of the coalition of sponsors.

It is clear that a necessary change in the organization usually causes pain. Because it could take them out of the organization, it could change their future in economic terms, it could affect their relationships—with colleagues, family, and friends—or even change the kind of work they like and want. Nothing can avoid the pain of loss. However, the respect, seriousness, and broad criteria with which the project is approached can change the emotions and the perception of the experience.

> *Sponsors' coalition responsibility: monitoring people's experience of change.*

I believe that it is the responsibility of the organization to act with full social responsibility. Those who design and implement the project, both in technical and social conception, must keep the employee experience as a basic principle. **Change management is not just about getting everyone on board.** It is about minimizing negative impacts, not only on the business and the project but also on those affected by the project.

> *Change management is not only for people to change, but for them to go through the change with the best possible experience.*

CHANGE IS AN ONGOING, CONTINUOUS EFFORT

In my experience, the first of Dr. W. E. Deming's 14 principles, Consistency of Purpose, is a key. If you choose one approache, don't think that it will solve all your problems, that you will get immediate results, or that it will work for every department, section, or person. Don't change your approach too quickly. There are no magic approaches. Perhaps the metrics I discuss later in this book will help you clear your mind along these lines.

> *Don't change your approach too quickly. There are no magic approaches.*

You should be suspicious of the initial support. When you need to meet goals—results, timelines and budget—don't be fooled by listening to people with the apparent coincidence of change. Everyone demands change, a director, a manager, an employee, an operator behind a machine, a salesperson. The problem lies in the fact that the change everyone is demanding is not the same and may not be the one proposed by the initiator and/or the one the organization needs.

> *Everyone demands change. The problem lies in the fact that change everyone is demanding is not the same.*

NOTES

1 Jeffrey K. Liker and David Meier, *Toyota Way Fieldbook, a Practical Guide for Implementing Toyota's 4Ps*. McGraw-Hill, 2006.
2 John P. Kotter, *Leading Change*. Harvard Business Review Press, 2012.
3 8D, Agile, Design Thinking, Lean, Project Management, Six Sigma DMADV, Six Sigma DMAIC.
4 Marcos E. J. Bertin. ASQ honorary member. IAQ honorary member.
5 https://asociaciondec.org. Association for the Development of Customer Experience.
6 www.jbs.cam.ac.uk/people/kamal-munir/.
7 www.st-edmunds.cam.ac.uk/people/professor-shahzad-ansari/.
8 ® ADKAR and/or Prosci is a registered trademark of Prosci, Inc. All rights reserved.

11 The Path

Raúl Molteni

WHAT DO I MEAN WHEN I SAY PATH

Plans usually do not turn out the way they are designed. The ability of leaders, sponsors, and the facilitation team to evaluate, learn, and improve the plans will be as important, if not more important, than the design itself.

IF PLANS DON'T WORK AS PLANNED, THEN WHY DO THE PLANS IN THE FIRST PLACE?

In a management meeting, the commercial director asked about planning in an environment of very high uncertainty and inflation above 100% per year. The president and CEO responded: "We will plan to have as many things planned as possible and to have time to think and act on unforeseen events." The journey along this path will require instruments. We'll see them as we develop each stage, but here is a preview:

- Understand the business scenario.
- Assess the organization's readiness and capacity for change.
- Assess the impact on all stakeholders.
- Define the strategy and the link between strategy and execution.
- Design a plan according to the expectations of the leaders and the real possibilities of the system, in a strict balance, and integration between the technical and the human.
- Design a rigorous metrics system for each and every area of intervention.
- Implement a model that defines the resources to be used according to need, understanding that "one size fits all" is neither the most efficient nor the most effective.

THE PATH

My proposal has a similar path to many technical and operational methodologies:

- *What is the aim?* What is the vision about? What behaviors are you trying to change?

 DOI: 10.1201/9781003544807-14

- *What is the baseline?* What is the current culture? What beliefs and behaviors characterize it? What is the history of change? What will be the consequences of implementing the changes? How prepared are you to implement the change? What data supports your analysis and conclusions?
- *What is the gap, and what are the key barriers to change?* And what are the key enablers? Which deserve attention and focus? What are the values and behaviors to be preserved? What are the values and behaviors that need to be changed? What are the values and behaviors that need to be replaced? What are the risks? What data supports the analysis and conclusions?
- *How will the change be managed?* What is the strategy and actions to be implemented to achieve the desired changes? What are the actions that will address the key barriers? How will the enablers be used?
- *What are the necessary changes to be installed?* What is the plan? What will be monitored to track the evolution of the adoption of the new behaviors? What data will support the changes and results?
- *How will plans be monitored* and continuity of practices and results be ensured? When and how will learning occur systematically? What data will be used?

THE CORE PROCESS

I have previously discussed alternative change management processes and acknowledge the benefits of their implementation. My proposed process differs in that it mirrors the fundamental logic typically applied in technical projects. My aim is to provide simplicity and ease through the use of common stages and terminology. I aim to demystify the roadmap and have tested it, as well as seen it tested by professionals without humanistic training.

My proposed process differs in that it mirrors the fundamental logic typically applied in technical projects.

Of course, the support of professionals with competencies linked to human resources will be necessary and indispensable in some cases. In the same way that when a Black Belt proposes a design of experiments, he relies on those who have more specific knowledge of the physical aspects of the variables that will be at stake. In the same way that a Scrum Master relies on the development team. It will not be a separate and parallel input to technical development. It will be an integration of knowledge to address aspects that, in practice, occur simultaneously:

The project proposes a change, which requires a behavior of someone used to do it differently, who says something about the change and about himself in front of the change, and that leads him to say things to himself, to feel emotions and to have judgments, not always favorable to the new situation. Which in turn leads to reactions and behaviors, which are finally seen by other people as resistance to change.

The proposed process has two main stages. The Preparation and Learning stage and the Design and Implementation stage. Clearly, it will be circular. It will "circulate" as many times throughout the project as necessary, fed by indicator reading and learning.

The steps of the process are illustrated in Figure 11.1 and are as follows:

1. Understanding the Vision and Scope of Change

What, why, for what, to whom, how, and by when? How does it relate to the organization's strategy, goals, and other plans?

We seek to understand the vision of the change and often find that phrases are sought to define it. My focus is more on understanding what will remain, what will change, and what should be added. We also need to understand who and what will change from the technical side, and what beliefs, paradigms, knowledge, behaviors, policies, and processes need to change.

What does "improve customer satisfaction" mean as a goal? It can mean a little or a lot, either to people close to the customer or far away. Which customers are involved? What aspects create satisfaction or dissatisfaction? Which products or services are being discussed? Knowing or estimating this at the beginning of the project can help anticipate the reaction it may provoke in people.

About the project, who is promoting it, why, for what? What happens if it is done, what happens if it is not done, and why should it be done now? What is its impact? On the workplace? On compensation, metrics, structure, relationships, roles and hierarchies, processes, tools and systems, and mindsets and attitudes?

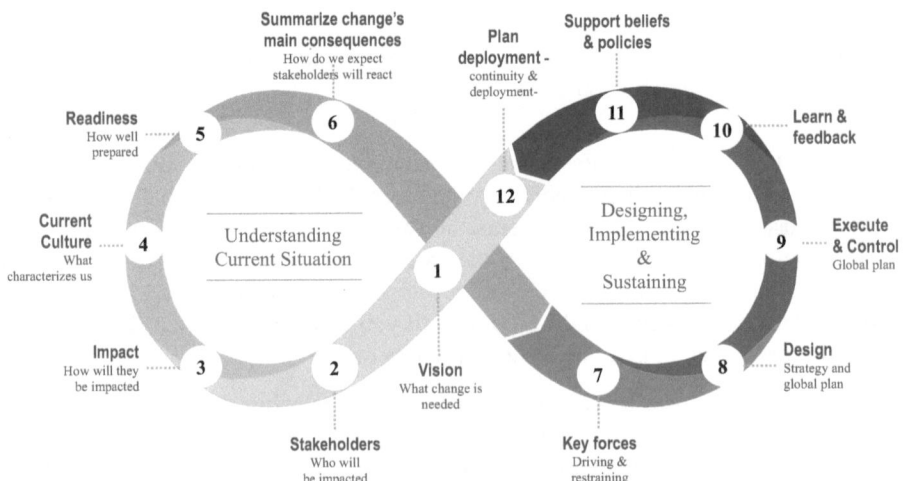

FIGURE 11.1 The main two steps for the execution of a project with a significant impact on the organization's culture: the deep understanding of the current situation, and the design, implementation, and sustaining of the technical and social results.

2. Identifying Stakeholders—Those Affected by the Change

Who is affected, internally and externally? Who will be affected by the defined change?

It's not just a matter of listing roles. A group of maintenance operators may perceive and experience the project differently from another group of maintenance operators, and in the same way as a group of Production operators. Segmenting groups is similar to what a Black Belt does when segmenting customers or problems in an improvement project. It has the same purpose: to better understand and focus on subsequent actions.

This will lead to an analysis of the impact of the change on each group, as a combination of culture, type of change, and how prepared they are will lead to an understanding of how they will be affected.

3. Appraising the Level of Impact—On Each Stakeholder

How will each stakeholder group be affected?

Having understood the change and the groups—not by function—that will be affected, it is now time to understand how the stakeholders will be affected; for example, what aspects of their daily lives will change, and what values, beliefs, behaviors, relationships, expectations, and emotions are likely to be different?

In addition, we seek to understand a particular stakeholder group: the sponsors, the promoters of change in the eyes of the stakeholders. How will they themselves be affected? For example, it is of interest to know whether they form a strong coalition and whether they have similar intentions and commitments to each other.

Steps 2 and 3 should be done simultaneously.

4. Characterizing the Current Culture—As It Is

What is the culture of the organization, what are its characteristics, and what is its background?

The leadership style, the level of interaction between people from different functions or people from different hierarchies, the habit and meaning of people working in teams, the history of similar changes, and the flexibility to face them are some of the characteristics of the culture that will be used to estimate how people might react to this new project and the proposed changes.

5. Assessing Readiness for Change

How well are they prepared for change? How well is the team, project, and organization prepared for change? Are all aspects—sponsorship, project management, change management, and technical—integrated? How well is this project and the change related to other projects, current or to be initiated?

It is the analysis of the readiness of the organization, the executive sponsor, each sponsor, and each stakeholder in terms of technical and cultural change. Of course, we may find that the level of readiness is minimal or zero, which implies that there is a need to install contingency plans before a kickoff or go-live.

6. Summarizing Key Implications of Change

What do we expect to happen?

What are the expected stakeholder reactions? What might be the blind spots—the things we don't know?

The key is to understand the behaviors and reactions that can be prevented and mitigated. It is about understanding what responses to change we can expect from each stakeholder group, how they will react, and why. It is also what we need to measure to be sure we are moving toward compliance with the new practices, habits, and behaviors.

7. Determining Key Forces—Driving and Restraining

What are the key aspects of the change?

It is about making explicit, clear, and visible the aspects of the change itself, as well as the culture that will act as enablers and inhibitors of the change.

8. Design Strategy and Global Plan—Prevention and Mitigation

What is the best strategy and the best comprehensive plan that can be developed with current knowledge?

It is the definition of how to approach the global situation, the axes and milestones of the change strategy, and an integral plan that takes into account all aspects, both technical and social. The aim is to prevent as many potential resistances as possible, to be prepared to deal with them when they arise, and to maximize the conditions that have emerged as favorable to change.

The integrated plan is a combination of programs, plans, and activities that cover social and technical aspects and are worth focusing on. It may include:

i.   Definition of the Project Facilitator team.
ii.   The Project Implementation Plan from a technical perspective—for the standardization, improvement, design, agility, digitization, transformation, or evolution.
iii. Preparation of the Global Plan for the project.
iv.   Preparation of the Organizational Alignment process.
v.   Preparation of the Communication Plan.
vi.   Preparation of the Mentoring and Coaching Plan.
vii. Preparation of the Education and Training Plan.
viii. Preparation of the Resistance Management Plan.
ix.   Creation and formalization of the Monitoring System.
x.   Preparation of Performance and Communication Guidelines for the different roles.

9. Implement and Monitor and Feedback Plan in Real Time

Execution of the plan in accordance with the rules of project management and PMI best practices.

10. Learn and Feedback Plan in Real Time

What works and what does not work in the developed plan? Under what conditions and criteria does it work or work not? What needs to be improved? What have we learned so far? How, when, and where can this learning be used? What knowledge and beliefs should we reconsider in the light of facts and data?

It is the analysis of the installed indicators and what is working according to plan and the results obtained. In addition, it is checking what is not working according to plan or what is producing results below what is expected or possible. It involves identifying and prioritizing the few programs, plans, or activities that need to be focused on to improve compliance with the changes.

11. Sustaining the Change

How to support all that has been achieved? How to ensure continuity and sustainability?

This includes recognition activities for those who have been involved, encouragement and involvement of others, and activities aimed at continuing and intensifying the review of policies, processes, procedures, resource allocation, and structures that may act as barriers.

12. Deployment Plan

What is the continuity plan? What is the plan for the deployment of concepts and behaviors? What actions can be eliminated because they do not serve the change? Who should act differently?

Part IV

In Preparation for the Project

This section outlines the preliminary steps required for the project. It discusses the roles of project leadership, which involves management, sponsors, and/or the coalition of sponsors leading the project in a strategic and visionary manner, as well as a facilitator team leading the project operationally. It also offers suggestions for the optimal allocation of these roles.

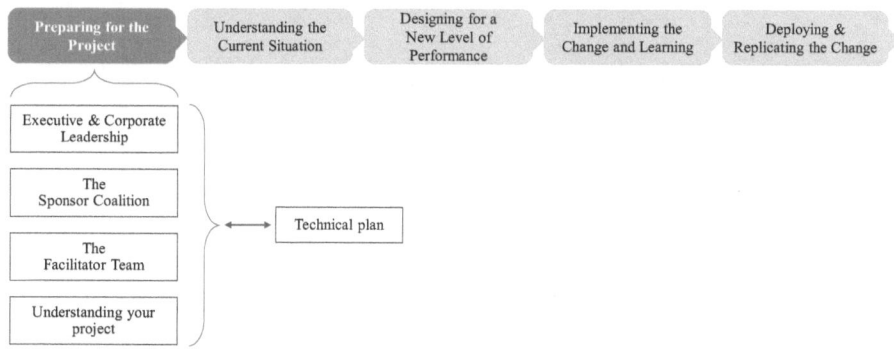

FIGURE 0.4 Four steps of the Preparing for the Project phase: ensuring executive and corporate leadership, forming the sponsor coalition, putting the facilitator team in place, and fully and deeply understanding your project. The last three steps are discussed in Chapters 13, 14, and 15.

DOI: 10.1201/9781003544807-15

12 Executive and Corporate Leadership

Raúl Molteni

THE PURPOSE OF EXECUTIVE LEADERSHIP

Establish project leadership with a sponsor, executive sponsor, or sponsors' coalition with full authority, background, and unity of purpose to "shoulder the project."

PROJECT LEADERSHIP

> *The big key to a good project and effective change is the leader's passion, commitment, and consistency of purpose.*

Executives, sponsors, managers, middle managers, and leaders—with their roles, attitudes, beliefs, and commitment—are the key to making the project stick. The responsibility of the facilitator's team is to design, execute, and control actions to support and sustain this leadership, and the responsibility of the executives—owners, directors, and the C-suite—is to guide and support the facilitator team.

When they prioritize its own hierarchical level over the needs of the facilitation team, it can have a detrimental impact on the project. It is not a question of who depends on whom. It is about who has the best understanding of what the project requires:

- Strategy, guidelines, and leadership visibility.
- Knowledge and technical and social skills of the facilitation team.

> *If the project is to achieve significant results and impact on the culture, the commitment of the owner, board, CEO, and sponsors must be visible, coherent, and sustainable.*

DOI: 10.1201/9781003544807-16

KEY CONTRIBUTIONS FROM THE BOARD AND OWNERS

For a project to have an impact on results, contributions are required from the owners and board of directors. This is mainly to:

- Ensure the project is *aligned* with the organization's vision and strategy.
- Secure *resources*.
- Provide *support to sponsors and the facilitation team* for project management, including definition, execution, and monitoring.
- Install *indicators* to monitor the project and its results.
- Consider *both perspectives:* How does the project impact the *business results*? How does the project impact the *culture* of the organization?

It is important to pay equal attention to the design of the culture as to the maintenance of equipment and processes.

VALIDATE CONSISTENCY OF PURPOSE

It is essential that senior executives maintain a consistent and clear sense of purpose. The constancy of purpose requires that the project be given priority over any difficulties that may arise. It should not be assumed that this is a passing trend. It should be approached as a component of a larger undertaking. Ensuring that projects remain aligned with the organization's strategy and objectives, as well as with one another, is a key aspect of maintaining consistency of purpose. To ensure that right decisions are made, the following questions could be asked:

- Is it your responsibility to make this decision?
- Are you maintaining consistency between your principles and values and the project? Are you being honest with yourself when committing your support?
- Have you considered all the information available, or are you being guided by the enthusiasm and suggestions of others?
- Are you clear about the role, dedication, contribution, and visibility that the project requires from you?
- If it doesn't work, will you take the position of learning, or will you consider it a failure and blame someone else (if the latter, quit now)?

ABOUT THE FACILITATOR TEAM

Coordinating and facilitating change requires the assignment of a critical resource: the team of professionals who will lead the change operationally: Black Belt or Master Black Belt, Scrum Master, the PMO, the Operational Excellence function, or a CTO; a team with the technical background necessary to understand the objectives, design the project, structure, organize, and execute the strategy and plans, and control and provide feedback on the project.

A team with the technical background to understand the objectives, design the project, structure, organize, and execute the strategy and plans, and monitor and provide feedback on the project.

As was discussed in the introductory chapter, experience has shown us that this operational leadership team must also possess the necessary skills to plan, execute, and control actions related to social aspects. These skills include the ability to gain buy-in, minimize the impact on people, reduce resistance, conflicts, and stress, and work on how the culture should be absorbed or modified as a consequence of the project. There are options:

- Integrate HR personnel with change management training into the project to work as a team with the technical team.
- Train the technical team in the change management process (see Parts IV through VII).
- Form a functional team. Given the continuity of change that the organization is likely to face, it would be worth thinking about prioritizing the formation of a functional group in HR or Talent.

This is an excellent illustration of organizational consistency and commitment. A lack of willingness to allocate the necessary resources to promote and structure the project effectively raises questions about its overall priority. It is the responsibility of organizational leaders who are driving the change to ensure that the right people are assigned, prepared, and available to take on these operational roles.

DELEGATION TO THE FACILITATOR TEAM

The facilitator team has been delegated the responsibility of assuming leadership and operational control of the project. For the team to be held accountable, three conditions must be met. Otherwise, no delegation will occur.

- They know what to achieve and do. Are they accountable for what they do?
- They know their performance level. Are they delivering and achieving?
- They are able to change their performance. Can they adapt and improve?

If the team does not meet these three requirements, there is no effective delegation to the team.

THE ROLE OF MANAGEMENT AND MIDDLE MANAGERS

In many organizations, middle management is overlooked. The focus is on the highest hierarchical and operational levels. The focus is on those who "must do" and those who "must lead." However, it is the middle managers whose day-to-day actions determine the awareness and commitment of the employees to the project. As mentioned in the *Toyota Way Fieldbook*, a team leader has three responsibilities: support the operation, promote change, and lead change.

> *Middle management needs to be considered and prepared for change.*

The role of leaders and middle managers is to guide, align, involve, and keep the internal fire alive so that each person can realize their potential in their area of responsibility. It is not to give orders or determine what each project should be and do. **The role is to create the environment for others to fulfill their role** and allow them to find their own internal motivators. It is to create and maintain, as Hugo Strachan[1] says, an environment in which each person can develop and grow.

I have drawn on Jeffrey Liker and David Meier[2] to highlight the requirements that executives and middle managers must meet:

- *Conviction* on the need, importance, and the best future situation of the project. I am not talking about repeating messages; I am talking about real conviction.
- Real *Commitment* to lead the project and its group.
- *Competencies*. Acceptance of training to acquire knowledge and skills to lead and accompany change.
- *Consistency* to identify and maintain the conditions necessary to advance and consolidate the project—policies, processes, procedures, and environmental conditions that require changes and the allocation of more or new resources.

I emphasize on the following responsibilities as well:

- *Co-create the vision of the project* or deploy it throughout the organization, not as a product of the inspiration of the moment but as a product of strategic analysis.
- *Create opportunities* that, aligned with the vision, translate the vision into concrete project terms and objectives for managers, middle managers, and operators.
- *Secure resources* for the project. It is unrealistic to think of a significant change project without a budget for training, preparation, experimentation, and learning.
- *Align structure*, positions, relationships, staffing, and resources with project intentions.
- *Learn* continuously.
- *Inform, listen, reflect*, identify obstacles, forget the ego, and learn by correcting and improving the plan. Listen to the opinions and needs of the organization.
- *Create a consistent environment*. Create relationships and forget "divide and rule." Eliminate identified barriers such as policies and processes.
- *Make consistent business decisions*. Reflect the beliefs, values, and goals that guide the project in daily and other decisions.

- *Reinforce positive behaviors.* Encourage learning over negative behaviors. Highlight those who demonstrate behaviors consistent with project intent and change.
- *Differentiate* those who comply from those who do not. Do not allow "nothing happens if nothing happens."
- *Go beyond* your area, sector, or function; work on the system because you have direct responsibility or because you believe you can influence who does it.

CONSISTENT DECISIONS

Inconsistent decisions wipe out the efforts of the promoters of change in a matter of seconds. If a person who has a proven track record of ignoring the behaviors and routines required by the project—for example, refusing to meet KPIs in his or her area, not adhering to sprints, or systematically rejecting suggestions from improvement teams—and change is recognized, rewarded, or promoted over others, it sends a very clear signal that what is being promoted is voluntary. And if it is voluntary, it is not "much to worry about."

REGARDING THEIR OWN TRAINING

Usually, the focus is on the training of practitioners: Black Belts, Scrum Masters, POs, and PMs. It is recommended that the board, senior management, and the sponsors' coalition be the first to undergo training. This should include the roles, tasks, and activities that they themselves will be responsible for executing to demonstrate and reinforce the continuity of the change, as well as how to evaluate the evolution of the project.

LEARNING TO LEARN

Owners, directors, executives, and managers must learn to learn.

Understanding why a particular action yielded the desired result is key to facilitate learning. This would enable to gain a deeper understanding of the relationship between "practice fulfilled and result achieved,"[3] allowing us to identify what can be adopted or redesigned for deployment to another process, product, sector, or industry. This is a form of first-order learning.

However, it is imperative that owners, directors, executives, and managers adopt a new and more profound approach to learning. The result is a consequence of the action, but this, in turn, is a consequence of the knowledge, experiences, values, beliefs, and theories of those who generated the analysis and made the decision. Second-order learning is focused on identifying which of these is not applicable or should be questioned in order to design future actions and situations.
This means:

1. *Evaluating the project's progress based on "technical" outcomes*, such as a reduction in errors or cycle times, the number of failures encountered during the initial stages of a new product's development, or increased sales following the launch of a new product
2. *Assessing the project's progress in terms of cultural outcomes*, such as the percentage of personnel adhering to new customer service protocols, or the satisfaction rate among employees with the five-minute pre-shift briefing.
3. *Understanding* the theories, beliefs, experiences, knowledge, and stances that influenced the decision-making process.
4. *Challenging* these beliefs, experiences, knowledge, and stances in light of the results.
5. *Suggesting new actions* based on this reflection.

NOTES

1   Hugo Strachan, IAQ Academician, QiGTT, Former President for Hewlett Packard Argentina.
2   Jeffrey K. Liker and David Meier, *Toyota Way Fieldbook, a Practical Guide for Implementing Toyota's 4Ps.* McGraw-Hill, 2006.
3   See Chapter 8, "About Owners, the Board, and CEOs."

13 The Sponsors' Coalition

Raúl Molteni

PURPOSE OF THE SPONSORS' COALITION

The assignment is to prepare a person or team with authority, who understands the project and has the passion to carry it forward despite difficulties. The team understands the need for training and learning as a means of identifying and addressing inhibitors throughout the project.

> *A good executive, a good sponsor, a good facilitator puts "what can I do" before "it's the fault of ... ."*

REGARDING THE SPONSOR

Not even the best group of facilitators or change promoters can replace the impact that a real **leader can have on people by communicating the excitement, need, and passion behind the project's purpose**. Neither can a video or a representative. It's the leader, period.

Directors and managers may think that just because they have done the kickoff, made a few presentations, and assigned a group of facilitators, their job is done, and everything should work. And if it doesn't work, they need to understand that they are the ones who failed.

Various authors emphasize the role and importance of the sponsor for a project. However, it is common to see surveys that show dissatisfaction with the fulfillment of the role. In the eleventh edition of the Prosci®Change Management Best Practices Benchmarking Research Report,[1] project teams report that half of their sponsors have little or no understanding of their role in managing the human side of change. And, according to the study, only 33% found their sponsors to be highly effective.

Our experience is consistent. The role of the sponsor, or sponsor and champion—as commonly found in Six Sigma projects—and that of the product owner—in Agile projects—are played by managers who act in front of project teams as they see fit, and not necessarily as they were assigned to do. **Sponsors are senior executives who have the authority, credibility, and resources to support and enable the change**.

DOI: 10.1201/9781003544807-17

In addition to their responsibility, power, and executive vision, sponsors must be highly respected individuals within the organization and have the understanding, enthusiasm, conviction, and sufficient passion for the project to lead and motivate.

WHY A SPONSOR

Projects that impact culture are decided, defined, and established by management, if not by owners and executives. However, it is common to hear them referred to as "so-and-so's Black Belt project" or "so-and-so's team project." If the project has an impact, it is likely to affect more than one area and/or require the contribution of people from different functions and levels. The project does not belong to its leader, it belongs to the organization.

Sponsors are people of such hierarchical level that they lend their authority, formality, credibility, and support to the structuring and development of the project. They have a key responsibility: to ensure that the project manager and team have the necessary skills and resources to develop the project. They are aligned with the strategic direction of the organization and the vision of the project, identify, understand, and communicate the benefits and consequences that the change will have on the organization. They become the "owner" of the change, taking responsibility for it, leading it.

The sponsor communicates with stakeholders in person and builds trust by demonstrating consistency between verbal and written communications; identifies the root causes of resistance to change and addresses them. He does not see resistance as a "negative" or regrettable occurrence, but rather as a natural and valuable source of information for improvement.

The sponsor implements and oversees initiatives that reinforce the change process. He provides recognition and identifies and promotes changes to the systems in place, maintaining an empathetic position at all times.

A good sponsor exemplifies the qualities of a good leader. The role entails guiding, supporting, and assisting by removing obstacles. It is based on the value of the question, not the correctness of the answer. The sponsor guides action, rather than determining it. He listens with understanding rather than dictating. He considers the other person's perspective rather than pass judgment.

KEY FEATURES OF A GOOD SPONSOR

A good sponsor should adopt a facilitative approach rather than a managerial one. His position within the organization and his experience afford him the ability to see the business and guide the teams toward the needs of the project and strategy. His positive attitude enables him to assist and collaborate with the team, removing potential obstacles that the project team(s) may face. His role is that of a service provider. **The role is not to demand results but to facilitate them**.

The sponsor clearly communicates the objectives, scope of the change, and expectations for each stakeholder. He ensures alignment between the change initiative and the business strategy, as well as the anticipated outcomes of the project. He identifies critical decisions and the appropriate stakeholders for involvement. He anticipates future developments and acts accordingly. He maintains a visible presence throughout the project. He monitors the resources required by the team and ensures they are available at the appropriate time.

THE ROLE AND RESPONSIBILITIES OF THE SPONSOR, THE EXECUTIVE SPONSOR, OR THE SPONSORS' COALITION

Depending on the size of the project, there may be one or more sponsors. If there are several, one may act as the executive sponsor, closest to facilitating and ensuring alignment and resources to the project teams. In terms of change management, the executive sponsor has six main challenges in his role:

- *Build coalitions with peers and senior management.* Achieve a critical mass of support that makes the project "something that matters to us and should be part of our day-to-day life" and is seen as "our" project, not "my" project.
- *Train and maintain a passionate and skilled facilitator team.* Assign members and ensure that the team has the necessary competencies. Maintain systematic, frequent, and personal contact with them to understand them, the project, and each other.
- *Participate actively and visibly throughout the project.* Communicate the whys and wherefores of change, the risks of not changing, and the sense of urgency. Maintain the validity and importance of the project throughout its duration.
- *Communicate personally with all stakeholders*, especially senior management. Understand and personally work with those who show signs of dissent or resistance. Understand what is needed to ensure that everyone has sufficient awareness, commitment, and skills. Emphasize organizational commitment. Resolve potential political and operational conflicts. Reinforce positive behaviors and help reflect on negative ones. Ensure that messages reach all employees in the organization and other stakeholders such as suppliers, customers, and the union.
- *Involve everyone.* Ensure that plans are designed and executed to include all those directly and indirectly affected by the project.
- *Be accountable for identifying, addressing, and eliminating inhibitors.* Yours is the main role that should take care of the organizational resistance. So, identifying those values, paradigms, policies, and processes inconsistent with the desired change and improving them is a must.

SPONSORS NEED TRAINING

Experience has shown that most individuals acting in the role of sponsor, with the function of guiding the project team, require training to perform their role effectively.

While reading a book and listening to other experiences can provide valuable insights, they are not a substitute for a training activity.[2] Rather, they complement it. Training offers a unique opportunity for understanding by bringing the theory to the "reality of each participant" and provides an environment for the exchange of ideas among the participants.

It is the responsibility of the facilitator team to provide support to the sponsors so that they can fulfill their leadership role in a clear and consistent manner. This includes preparing them to meet their responsibilities and establishing the criteria required by the project. The objective is to ensure continuity in the reflection process: what is the project and its participants looking for from me?

WHAT IS A COALITION OF SPONSORS?

A network of support. A "single voice." Strategic allies working in a network to support and lead the project. Alliances that must be seen both horizontally and vertically; between leaders, as well as between the manager, the boss, and the supervisor of a given sector. Leadership coordination and alignment are key factors in getting people on board with change.

> *Strategic allies working in a network to support and lead the project.*

The coalition of sponsors is absolutely necessary when changes go beyond one management or one area. Different messages and positions from leaders of different areas in the position to execute change send signals that there is really no change, just a change that everyone can pretend or ignore. The coalition attacks situations such as "there are managers who do not agree with the change" or "the project belongs to the manager of ... ." It prefers to communicate the sense of the group to that of a "nonconsensual project."

A characteristic of a coalition, and not of a group, is that sincere dialog prevails over tensions, unspoken conflicts, personal interests, and intentions to highlight oneself at the expense of others. All the rules of teamwork apply to this group. For example, maintaining a shared vision and using data. However, the key is how they manage the intention to excel over others to be better positioned for future recognition, compensation, and promotions.

> *The key is how they manage the intention to stand out from others to be better positioned for future recognition, compensation, and promotion.*

They will most likely need support from the facilitator team to move from a "I win because I know better" stance to a "win-win" stance. The rules of consensus, starting with "first the coincidences, then and only then the differences," should be used routinely by the group.

A Good Coalition Complies with the Following Characteristics:

- It comprises members who are willing to invest the time and resources necessary to make the project a reality. Furthermore, it can influence and ensure compliance within the organization.
- It is aligned with the diversity of groups, including geographical areas, functional areas, and personnel segments, that will be affected.
- It recognizes the need for assistance and relies on the facilitator team. They foster interrelationships and avoid the "divide and conquer" strategy.
- It seeks and encourages the commitment of other executives to support change.
- It uses naïve listening, questioning, communication, and support to keep the facilitator team and the rest of the organization active and engaged.
- It looks for opportunities to find small, quick successes to maintain motivation.
- It proactively looks for inhibitors and generates or promotes actions to address them, especially organizational ones. These actions reinforce positive behaviors.

As Liza Kempton says, "Simply put, people will not usually be more committed to change than the sponsor."[3]

A coalition is a convergence of forces and interests that are unified by a common objective and willing to provide support, even if there are discrepancies in opinion. Consensus does not imply total agreement but rather a commitment to back a decision or a position.

It is an ongoing effort that must be led by the executive sponsor and methodically supported by the facilitation team. Frequent communication and learning activities are essential. It is also important to assess the role and performance of the sponsor-both as a coalition and as individually. This is the example to which the rest of the organization will refer.

If clarification is needed, the owners, board, and CEO must be part of the coalition or act and be fully aligned with it.

GUIDE FOR SPONSORS

It is a checklist and agenda that describes common or typical actions of the sponsor role and acts as an element of help and support for them. It enables them to perform their leadership role in a clear and consistent way. It gives them visibility and foresight into what the project needs from them and what actions they need to take. It helps to ensure that the sponsor is active and visible throughout the project, communicating and working in partnership with their colleagues; therefore, the document needs to be updated throughout the project. A guide for each sponsor is desirable. This guide will identify:

- The group or person of interest to approach or interact with.
- The activity to be performed.

- The goals/purpose of each activity.
- Details such as key messages, responses to be obtained, reflections to be shared.
- Dates and frequency.
- Comments or clarification of what the sponsor is trying to accomplish.

It may include activities that are present in other plans,[4] but it is intended to serve as a guide for the sponsor and allow him or her to consolidate all the activities to be carried out.

SUGGESTIONS

- Propose and offer this document to the sponsor as a support element.
- Establish a fluid back-and-forth activity with the sponsor to tailor this guide to their availability and agenda, in addition to the needs of the project.

BUILD AND MAINTAIN THE COALITION OF SPONSORS

The responsibility for the clear definition of the following issues should be shared between the executives—who are best positioned to understand the objectives of the project and change—the facilitator team—who are best prepared to anticipate the impact of the change—and the sponsor of the organization.

1. Determine which sponsor or sponsors should comprise the coalition of sponsors.
2. Prepare the sponsor or coalition to understand the process to be followed and the role they should assume. Give them training and, in general, additional support to reflect on how they evolve.
3. Build with them the "Guide for Sponsors."
4. Take the time to reflect and find consensus on the project's objectives, process, risks, and key success factors. Define the roles to be assumed.
5. Define and install a comprehensive project dashboard (see indicators in Chapter 27, "Metrics and KPIs").
6. Arrange regular meetings to analyze the evolution of the project and how each sponsor and the coalition is performing.
7. Implement the necessary changes in the training, roles, individual contributions, and the plans in which they are involved.

NOTES

1 Lisa Kempton, www.prosci.com/es/blog/5-maneras-de-ayudar-a-los-patrocinadores-a-construir-una-Coalición-de-apoyo-para-el-cambio.
2 Raúl Molteni and Oscar Cecchi, *Lean Six Sigma Leadership*. Ediciones Macchi, 2005.
3 Kempton, www.prosci.com/es/blog/5-maneras-de-ayudar-a-los-patrocinadores-a-construir-una-Coalición-de-apoyo-para-el-cambio.
4 See the tools in Part VI.

14 The Facilitator Team

Raúl Molteni

THE PURPOSE OF THE FACILITATOR TEAM

An operational project leadership team, aligned with the sponsors, trained and resourced to develop, coordinate, monitor, and improve a comprehensive change project.

THE FACILITATOR TEAM THAT COORDINATES THE CHANGE

It is the responsibility of the change promoters, owner, board, CEO, director, or manager of the business unit, depending on the entity over which the project has scope, to ensure the designation, formation, and operation of this facilitator group.

UNIFIED TEAM

The coordination and facilitation of change require a combination of technical and operational skills. A team with the technical background to understand objectives, design the project, structure, organize, and execute strategy and plans, and control and provide feedback to the project—Black Belt or Master Black Belt, Scrum Master, the PMO, the Operational Excellence function, or a CTO—is essential.

In addition, planning, executing, and controlling actions that address social aspects is also important. It should have the competencies to gain the support of individuals and to reduce the impact on people, conflicts, and stress. It is crucial to understand how the culture can be adapted to accommodate the project's outcomes. In light of the ongoing changes the organization may encounter, it is advantageous to prioritize the formation of a functional group in the areas of human resources or talent management. One potential solution is the establishment of a change management office.

Nevertheless, our experience has shown that having two separate groups—one with technical and the other with social specialties—can lead to a loss of integrity. In addition, a functional change management group, as recommended by various "best practices," may be prone to addressing change for the sake of change.

DOI: 10.1201/9781003544807-18

> *The change must have a strategic or operational origin that is aimed at the evolution of the organization with concrete results in terms of positioning, product, or service. The objective is not simply to implement change for its own sake but rather to achieve tangible benefits for the business.*

The individual or group with technical expertise and the individual or group with training in the social aspects must collaborate to develop a unified project plan. They must also structure and coordinate their work with the rest of the functional areas. It is not the responsibility of the personnel in operational areas who are implementing changes to understand that the discussions of change management and process improvement by different functional units are part of the same overall process. The staff view the two projects as distinct.

This synchronization is part of the sponsors' coalition responsibility. It is essential to create and allocate sufficient resources to a facilitating and coordinating group that, as a team, has all the necessary competencies.

> *It is the responsibility of the sponsors' coalition to ensure that a facilitator and coordinating group are provided with sufficient resources.*

WHO SHOULD BE PART OF THE FACILITATOR TEAM

The facilitator team should include individuals with competencies related to social aspects, such as human resources (HR). These professionals can make significant contributions in terms of culture awareness, communication plans, training, coaching, and resistance management. Individuals with project management (PM) competencies will be able to contribute more effectively in the areas of plan generation, structure, follow-up, and feedback. Those with technical competencies related to the project—Black Belts, Green Belts, Scrum Masters, Lean Practitioners, and Quality or Operation personnel—will have a greater role in defining activities of this type.

Each of them will contribute to observing how the intentions and objectives set forth by their respective competencies affect and influence the objectives and intentions of the others. They will also have the chance to collaborate with one another to ensure that the project evolves with minimal resistance, inconveniences, and risks.

PREREQUISITE: EMPATHY

This refers to their ability to identify with all stakeholders who, in one way or another, are impacted by the change.

PROFILE OF THE CHANGE FACILITATOR

I consider the following to be some of the characteristics and competencies that are needed in the team. This means that not all of them need to be found in every facilitator. It is the complementarity between them that gives strength to the team. On the contrary, these competencies could be acquired through training, and the predisposition to them and the personal characteristics are the criteria that should be used for the selection of the team members.

PERSONAL SKILLS

- Integrity.
- Strategic vision.
- Quest for personal development. Capacity for "inveterate" learning.
- Customer-oriented approach.
- Self-awareness and self-control.
- Initiative and autonomy. Persistence and resilience. Resistance to frustration.
- Emotional intelligence.

SOCIAL SKILLS

- Understanding the human dynamics of change and the change process.
- Ability to communicate effectively, inspire, and encourage. Empathy and 360° communication.
- Effective meeting and team facilitation skills.
- Conflict management.
- Building value-based relationships.

TECHNICAL SKILLS

- Understanding the business and the context in which it operates.
- Understanding the concepts, principles, and tools of the technical methodology to be used.
- Problem analysis using data and methodologies.
- Ability to plan and negotiate with a win-win vision.
- Creativity techniques.

FACILITATOR GROUP RESPONSIBILITIES

The facilitator group has several responsibilities throughout the project that should be continuously analyzed and monitored by the executive sponsor or coalition of sponsors:

- *The design of the project, its strategy, and change plan.* I refer to the overall project and methodology. It includes the overall planning, considering the phases of initial situation assessment, strategy and plan design, analysis and identification of key causes and opportunities, generation of solutions that integrate all perspectives through a global plan, and control of the sustainability of results and practices.

- The design, *implementation*, and effectiveness of the activities included in the plan.
- *Supervising and modeling the methodology to be applied*—Lean, Six Sigma, Agile, Transformation, Digitalization, Designs, or other equivalents.
- *Aligning, communicating, training, supporting* individuals and groups, and managing resistance according to the global plan. By themselves or through third parties.
- *Controlling the execution of the plan.* Effectiveness—quality, time and costs; maintaining the necessary indicators to analyze progress; systematically and frequently analyzing the evolution of the project with the sponsor or sponsors' coalition.
- *Learning.* It is essential to systematically and frequently analyze to learn and give feedback on the strategy and plan, suggest improvements to the project, and use results of all kinds—such as those from indicators, training, coaching, individual and group feedback from employees and customers—to review progress and to design, propose, promote, and implement changes.

PREPARING THE FACILITATOR TEAM

1. *Define Sponsorship.* Depending on the scope of the project, a board member, owner, CEO, or functional unit manager should analyze and assign a sponsor, executive sponsor, and/or a sponsors' coalition.
2. *Determine if there will be a Change Management Office.* This may be a good option if frequent changes with a high impact on the organization are anticipated.
3. *Structure the facilitator team.* Include members who understand and coordinate the technical project, as well as change management and project management specialists.
4. *Train them in the competencies* that are not the "original" of each of them. Achieve an integral understanding of the project and its impact on the whole group.
5. *Create the Responsibility Matrix*—RACI (Responsible, Accountable, Consulted, Informed)—to clarify the purpose and responsibilities.

OPERATION

It will be necessary to build the team, equip them, and imbue them with the project's mystique in addition to its "purpose."

15 Understanding Your Project

Raúl Molteni

UNDERSTANDING THE PROJECT AND CHANGE

The Change Canvas is the physical result of this phase. It reflects the conclusions about the meaning, circumstances, difficulties, and benefits that result from the leaders' and sponsors' understanding of the true purpose of the project.

> *Strategic positioning is impossible without strategic thinking. Vision is not about what can be done, it's about what must be done.*

ABOUT THE VISION

It is the future picture of what you want to create and achieve, described in the present tense. The vision of change sets the direction, the True North, to head. The key is not to think it is anchored in today, as that positioning takes away freedom and design options, limits possibilities, and leads to "because it can't be done" instead of "what if we could ... ." "Imagination and inspiration are important parts of a vision."[1]

> *It is better to have a clear vision and a sound plan and accept the possibility of failure than to have neither and remove all doubt.*[2]

Without a vision, you run the risk of designing and doing things that are evaluated independently and by different criteria. As a result, good ideas are implemented instead of good solutions, or resources—budgets and people—are allocated according to the priority of the moment or enthusiasm for a project. As Mike Rother notes in his book *Toyota Kata*, "It may be that our production true north is theoretical and not achievable, but that does not matter. For us, it serves as a direction giver, and we do not spend time discussing whether or not it is achievable. We do spend a lot of effort trying to move closer to it."

DOI: 10.1201/9781003544807-19

> *Without a vision, you run the risk of designing and doing things that are evaluated independently and by different criteria.*

How to Communicate the Vision

A good vision should be something that can be communicated in less than three minutes. The elevator pitch is a very good way to do this. It is common to find organizations that state the vision clearly but without further analysis and understanding of what they are really trying to achieve. The time and effort are put into achieving a clarifying and motivating phrase rather than understanding the type of change and future situation that is needed and desired. Expressions such as "Being an Agile Company" are confusing because everyone thinks they understand without the slightest agreement.

> *Expressions such as "Being an Agile Company" are confusing because everyone thinks they understand without the slightest agreement.*

AMBITIOUS VISION FOR THE FUTURE, BUT NOT NECESSARILY GLOBAL

Vision has to do with passion and conviction. Passion and conviction are driven by the future, by strategic thinking. It is the pursuit of a medium- and long-term goal when the logic of the present indicates that it is absurd. Vision is fueled by the logic of the future, not the logic of the present.

> *It is to pursue a medium- and long-term goal when current logic indicates that it is absurd.*

The vision does not necessarily have to be the big challenge or have the entire organization as its scope. It can be a narrow opportunity and still result from thinking big. "There are always pressures to be sure an opportunity is big enough, but most really big solutions began small and built momentum."[3] It is important that the vision is aligned with the strategy, as this will guide decision-making and coordinated and integrated action. The journey will be one of continuous improvement based on learning, an "unclear territory by being sensitive to and responding to actual conditions on the ground."[4] As Alvin Toffler[5] stated, "You've got to think about big things while you're doing small things, so that small things go in the right direction."

It is a mistake to think that a project is of greater scope or importance to the organization than it really is. It will inevitably lead to disappointment. In their intention to support a manager, management may agree to give a project more organizational space than it really has. They need to know that even if their intentions are good, they are creating serious problems of credibility and future results. The project must be given the space it deserves.

It is a mistake to think that a project is of greater scope or importance to the organization than it really is. It will inevitably lead to disappointment.

What should be done is to articulate, intertwine, and integrate projects to give them organizational meaning rather than unit, management, or sectoral meaning.

SHARED VISION

A good vision, with some exceptions, is arrived at after several moments of thinking, reflecting, exchanging ideas, and consolidating. The most interesting thing is that during this period, it is shared with different members of the organization. It is to be recreated and gain adherence and understanding at the moment of its public presentation.

Depending on the culture of the organization, the vision could be imposed, "sold," or tested by the opinions of others. Or, it could be the result of co-creation. But in all cases, a good project and change vision should:

- Be sincere and honest.
- Be simple, be clear about what is already defined and what should be cleared up and defined in the future.
- Emphasize the benefits to all stakeholders.
- Allow for the identification of concrete facts to be pursued as intermediate milestones.

"I am glad we didn't know that what we were doing was impossible."[6]

THE INSPIRATIONAL PHRASE IS NECESSARY BUT NOT SUFFICIENT

In impact projects, the vision has a technical component: design, standardization, improvement, innovation, agility, or transformation of operations across the value chain. Typically, the Project Charter reflects the technical perspective of the change. It also has a social component that reflects, for example, the values, beliefs, behaviors, practices, and habits that need to be challenged.

The question is: How will it change people's current work, feelings, and relationships? If the change, whether technical or cultural, is not understood in terms of the practices and behaviors to be installed, the impact on the way you operate, and the potential resistance that will arise, the likelihood of an "empty" vision is high. Understanding the vision means understanding its focus and consequences.

It is time well spent in conversation and reflection with those promoting change to understand, more clearly, the behaviors, practices, and habits that need to be changed, forgotten, or created. I know that it is impossible to do this with accuracy for three reasons.

- The technical aspiration may not yet be clearly defined.
- Reality will change and impose its share of uncertainty so that something planned today may not be the most convenient when the project moves forward.
- Very few executives initiate significant change with complete clarity about what they want and will need to change.

> *The question is: How are you going to change your current work, emotions, and relationships with people?*

SUGGESTIONS FOR UNDERSTANDING THE PROJECT, VISION, BEHAVIORS, AND IMPACT

INTERVIEWS WITH THE PROMOTER OF THE CHANGE

The greater the impact of the change on the organization, the higher the level of the interviewee should be. The promoter of change and the executive sponsor—if they are not the same person—may be the CEO, the director, the owners, or the entire board, depending on the nature and scope of the change. It is a good practice to make sure that the one giving the opinion and reflection is the true promoter of the project, the one who really has the intention and the greatest clarity of the change needed.

This is not an interview for the promoter or sponsor to answer questions; these are meetings—note the plural—to reflect on the change. These are the aspects that would be worth discussing with the promoter of the change.

1. About Business Impact and Subsequent Communication
    - *What we want to happen.* If we woke up tomorrow and the change was successful, what would we see? What would we not see? How will the organization see itself when this project is over?
    - *Why the change.* Why should we change? Why now? Why is it important to the strategy and the business?
    - *What is the change for?* What are the main objectives? What will it achieve? How will it impact the strategy and the business? How will it improve competitiveness? How will it affect processes? In terms of technology, how will it impact change or how will change impact technology? How will it impact the economic and financial results?
    - *Risks of not implementing the change:* What will happen if it is not done? What will be the impact? What will be lost?
    - *The scope of the change.* Are the aspects of the business to be affected concentrated—units, functions, processes, products, and services—or dispersed?

2. About the Results
    - *Results*. What results are expected and by when?
    - *Success metrics*. What are the indicators that would confirm that the change has been successful?
3. About the Organization
    - *Impact on the organization*. What organizational beliefs, policies, systems, processes, and procedures could be seen as somehow inconsistent with the change? And consistent?
    - *About other projects*. What other similar projects have been developed in recent years? How many and what other projects are underway in the organization today? Which could be added? What impact might they have—past, present, and future—on this new one?
4. About Stakeholders
    - *Stakeholders:* Who are the stakeholders who will be affected? Do they belong to a particular sector? Do they share anything in common conditions?
    - *Benefit:* For each of the stakeholders, what is the benefit of changing and what is the benefit of not changing?
    - *Impact on each stakeholder:* What changes will they experience? What new values, beliefs, or behaviors will they need to follow? What values, beliefs, or behaviors will they have to forget? How many people are in the group?
    - *Special conditions of the tasks:* What are the changes in terms of tasks, technology, and work environment? Are radical changes expected?
    - *Impact on hierarchies:* Are changes expected to alter the hierarchical status quo in the group? Will there be demotions or job eliminations? Or loss of aspirations for growth?
    - *Sponsor-related*. How does the group and the group leader perceive the project sponsor?
    - *On other projects:* What other similar projects have you developed in recent years? How many projects are going on today, and what other projects affect this group? Which could be added? What impact might they have—past, present, and future—on this new project?

Defining the vision with a phrase is still valid. These questions do not replace it; they serve to reflect and really understand what the organization or part of it is embarking on.

The answers to the above questions will then allow you to determine the impact of the change on the culture and operations. Any question that cannot be answered is an opportunity to better understand the vision, the project, and the resulting change. The answers will form the basis of concepts that will shape the key messages. Those, in turn, will drive and reference communication throughout the project. It will serve as a reference to determine the gap between the current culture and the one desired by the organization (see Figures 15.1a and 15.1b).

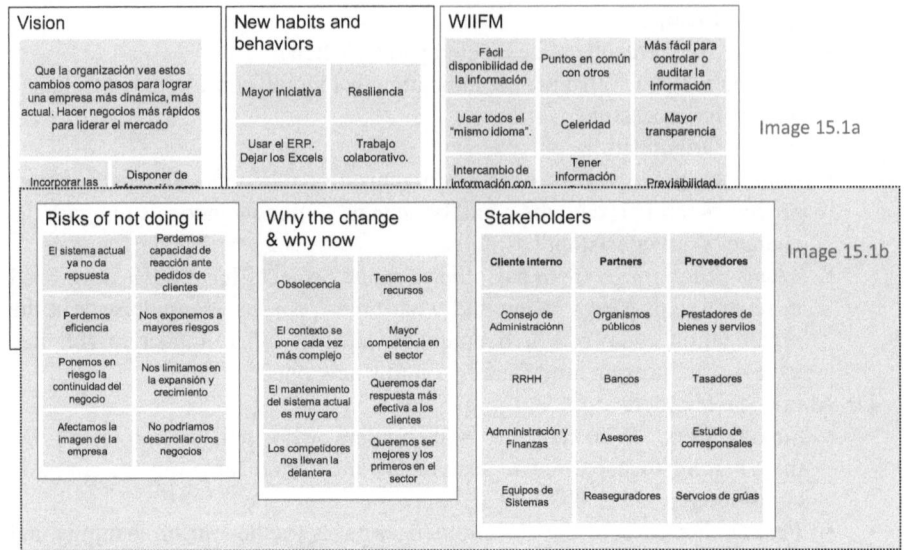

FIGURE 15.1A AND FIGURE 15.1B Sentence in which the management of a service company defines its vision and benefits, the risks of not achieving the vision, the reasons for undertaking the project at this time, the stakeholders and how they will be affected, and the description of the habits and behaviors that need to be changed.

NOTES

1 M. Joseph Juran and Blanton A. Godfrey, *Juran's Quality Handbook*, McGraw-Hill, 1999.
2 Bill Willcox. *The Imagineering Way*. Disney Editions. 2003.
3 Jeanne Liedtka and Tim Ogilvie, *Designing for Growth, a Design Thinking Tool Kit for Managers*. Columbia Business School, 2011.
4 Mike Rother, *Toyota Kata, Managing People for Improvement, Adaptiveness, and Superior Results*. McGraw-Hill, 2010.
5 [RM: Mike Rother, *Toyota Kata, Managing People for Improvement, Adaptiveness, and Superior Results*. McGraw-Hill, 2010]
6 Bob Gurr, "Original Imagineer." *The Imagineering Way,* Disney Editions, 2003.

Part V

Assessing by Understanding

This section outlines the second phase of the roadmap, which involves understanding the preliminary circumstances. The objective of the analysis is to gain insight into the following aspects related to the project in question: the project's current context, the current organizational culture, the level of readiness for change, the potential impact of the change on the organizational culture, and the implications for the project and the individuals involved.

The deliverable of this phase is to gain insight into the potential consequences of the project and change within the existing organizational structure in alignment with the previously established vision.

DOI: 10.1201/9781003544807-20

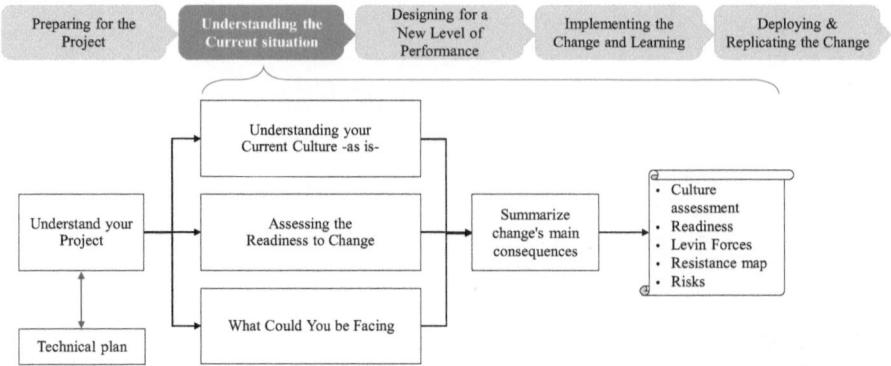

FIGURE 0.5   Three steps of the Understanding the Current Situation phase: Understanding Your Current Culture, Assessing the Readiness to Change, and What Could You Be Facing. The last two steps are discussed in Chapters 17 and 18.

16 Understanding Your Current Culture

Raúl Molteni

WHY CHARACTERIZE THE CULTURE

It is important to understand the degree of consistency between the current value system and culture, and to identify which values and characteristics should remain unchanged in order to facilitate lasting change with the project. By understanding the differences, it is possible to infer reactions that may favor or inhibit change, which in turn will inform the development of change management plans.

> *By understanding the culture, it will be possible to anticipate stakeholders' reactions to the intended changes.*

WHAT IS CULTURE

Authors such as Edgar Schein (2004) define organizational culture as "the set of beliefs shared by the members of an organization about the best way of doing things, which define the company's vision of itself and its environment." According to John P. Kotter,[1] culture refers to norms of behavior and shared values among a group of people. Norms of behavior are common or pervasive ways of acting that are found in a group and that persist because group members tend to teach these practices to new members, rewarding those who fit in and sanctioning those who do not. Culture, in a sense, binds the group together and gives it identity. To put it in everyday terms, culture is to a group what personality is to an individual.

> *Culture unites and gives identity to the group.*

Usually, we talk about the work environment. However, it is only part of the culture. Organizational culture is more than how people interact internally.

When discussing Agile, Six Sigma, or Lean methodologies, the consensus is that a "cultural shift" is necessary. John P. Kotter adds:

> Culture is important because it can powerfully influence human behavior. When the new practices made in a transformation effort are not compatible with the relevant cultures, they will always be subject to regression. Changes in a workgroup, a division, or an entire company can come undone, even after

DOI: 10.1201/9781003544807-21

years of effort, because the new approaches haven't been anchored firmly in group norms and values.

Every organization has a unique culture that distinguishes it, identifies it, **guides its actions, and governs its perceptions and the image it has of itself**. Each culture is distinctive, although some organizations have some common characteristics.

The culture defines and drives behaviors, influencing how people feel, act, and interpret changes, challenges, and opportunities in their environment. Behaviors affect key business indicators.

> *The collective beliefs, assumptions, habits, routines, protocols, and rules—both written and unwritten—that shape the behaviors of individuals within an organization contribute to the formation of its unique "personality."*

WHY IT IS CALLED INVISIBLE

It is called invisible because, unlike a device, a product, or a computer, culture is not material. It is like the air around us; we know it exists because it blows up a balloon, fills our lungs, ruffles our hair, or makes a feather fly. And just as we sometimes say "the air is heavy" without seeing it—for example, when the humidity and the temperature are high—we notice the characteristics of culture without seeing it.

YET, WE NOTICE IT

Although it is something invisible, culture drives behavior. When something happens—such as a speech from the CEO, the revelation of a fact, an incident, a result, or a customer complaint—it triggers internal messages and emotions in people and, consequently, influences their behavior in response.

> *Culture is important because although it is something invisible, it drives the behavior of people—owners, managers, supervisors, employees, and even suppliers—when something happens.*

In some cultures, someone says something in a meeting, and everyone falls silent and immediately looks at the most senior person. What the boss says is not questioned. Culture is not visible, but its effect on our behavior is. We distinguish culture by what it produces. By how we express ourselves, dress, act, and relate to each other. By how we speak and listen. By the condition of our desks, machines, bathrooms, dining rooms, and locker rooms. By spaces that are open to interaction or closed off by doors and walls.

> *Understanding this is key to predicting how changes will affect different groups.*

In response to a trigger, our internal messages and emotions, shaped by our cultural context, influence our actions. When an individual is asked to make a change, their previous experiences and beliefs influence the internal messages and emotions they experience in response to the request. These private messages often lead us to form judgments about what we are being told:

- "We've seen it before, and it doesn't work."
- "They ask us to change, but they don't change anything."
- "What's in it for me?"
- "Will I be able to do that?"
- "They think we're Toyota!"
- "All because we lost a few sales for two months."
- "Another fad!"

Some effects can be positive. A customer complaint can trigger feelings of "wounded pride" and provoke action to correct and save the situation, or negative reaction: "It's just one customer out of a thousand who complains."

Because it is invisible, we do not realize that it conditions our actions. Moreover, we unwittingly become teachers of the culture and unconsciously transmit it to those who join it. Meanwhile, we "see" the culture of organizations that are very different from our own. We listen to someone from Google, from Apple, from Amazon, from Toyota, from Lego, or from Spotify, and we "see" different things. We have seen, and often see, that those who admire some of the characteristics of these organizations deny and resist incorporating them into their organizations with various arguments.

Because it is invisible, we do not realize that it conditions our actions.

If we were all rational beings, the logic used to promote change would work. But we are also emotional, and as such, our beliefs and the perception left by previous experiences can lead us to resist change.

RESISTANCE IS NOT THAT BAD

As Cheryl M. Jekiel[2] puts it, "Lean efforts have failed in the past because the process was not approached as a complete cultural change." Most failures of standardization, improvement, and redesign projects stem from a failure to properly or seriously address the social impact of change.

It is worth returning to one concept: resistance is valuable for improving the project. De Bono[3] emphasizes the value of the Black Hat in certain circumstances. The arguments of those who disagree with the change or the action to be taken should be used to improve our plan, regardless of the emotions it generates. Let us use the arguments against us to identify the flaws in our plans. Luc Mayrand[4] says, "Guests' reactions to the experience will be instinctive and emotional, we need to tap into that too." Let us listen carefully to those who do not think as we do, especially to understand what we can improve.

> *Resistance is valuable for improving the project.*

In *The Fifth Discipline in Practice*, Peter Senge[5] clarifies: "The culture of an organization is constituted by the collective mental models of its members, so it is impossible to change an organization without examining its cultural assumptions." As Schein[6] states, "Culture is an abstraction, yet the forces that are created in social and organizational situations that derive from culture are powerful."

> *As Schein[7] states, "Culture is an abstraction, yet the forces that are created in social and organizational situations that derive from culture are powerful."*

WHY UNDERSTAND CULTURE?

To go back to Schein:

> If we understand the dynamics of culture, we will be less likely to be puzzled, irritated, and anxious when we encounter the unfamiliar and seemingly irrational behavior of people in organizations, and we will have a deeper understanding not only of why various groups of people or organizations can be so different but also why it is so hard to change them.

> *"If we understand the dynamics of culture, we will be less likely to be puzzled, irritated, and anxious when we encounter the unfamiliar and seemingly irrational behavior of people in organizations."*

We should take the time and effort to infer and simulate, as organically as possible, what will happen in the organization when the changes take place. "Where there is a disconnect in authenticity, people will feel betrayed and react negatively," says Tai Tran as cited by Simon Clatworthy.[8] Either changes that align with the existing culture are implemented or efforts are made to reinforce the culture to integrate the new beliefs, values, and behaviors associated with the changes.

While we cannot see it, we can characterize a culture by paying attention to how it manifests itself. This allows us to anticipate reactions to the project and the change we propose.

> *"People actually don't resist change; people resist being changed," Peter Scholtes.[9]>*

CRITERIA AND DIMENSIONS FOR CHARACTERIZING CULTURE

To understand and characterize culture, we propose the following model. We use eight dimensions that are relevant to the proposed changes in general. These dimensions are: (1) Leadership; (2) Decision-Making; (3) Autonomy, Collaboration, and Engagement; (4) Focus on Customers; (5) Improvement and Change; (6) Technology and Innovation; (7) Agility; and (8) Reinforcing Behaviors (see Table 16.1). The dimensions and the relative weighting can be adapted to the specifics of the project.

The intent is to anticipate enablers and inhibitory responses to change. These dimensions and the way they are addressed may raise serious questions. However, they make it possible to gain sufficient and reliable understanding to be effective in terms of time and cost. It is enough to:

- Anticipate and respond to resistance, projected and validated.
- Identify cultural conditions that are incompatible with the values, beliefs, and practices proposed by the project.

SURVEYING METHODS

In the case of projects requiring a global mindset shift for the organization, we recommend the following:

- *Meetings.* These have to be held with representatives from all levels of the organization, with a duration of between 30 and 45 minutes. The responses from staff will assist in the assignment of values as outlined in the table.
- *Surveys of all personnel. Surveying* using carefully chosen questions enables a broader view with comparisons and various correlations.
- *Touring the facilities*—get on the scene. Observing the staff at work and getting to know the condition of the common areas—reception, restrooms, locker rooms, hallways, dining room, and kitchen—give a very good picture of the culture.
- *Workshops.* It is an excellent alternative to develop workshops with representatives from all areas to gain deeper insights into specific aspects.

To obtain relevant information, it is advisable to also analyze other sources of information, such as:

- The union's vision, which can be a key to the success of the project.
- Internal communications.
- The climate survey, particularly, the level of teamwork, alignment with the organization, and satisfaction and relationship with the direct supervisor.
- Customer satisfaction surveys.

The union's vision could be the key to the project.

TABLE 16.1

The table displays the variables for each dimension that can be used to assess the current culture. Also shows the detail of what should be understand to assess, and how to score each of the variables

Variable	Scoring with 1	Scoring with 5	What we want to understand; what it is useful for
1. Leadership			
1.1 Shared values	The values are not known.	The values have been formally communicated and are understood by all relevant parties.	(i) If the values are aligned with the planned change. (ii) What preparation or work is needed to communicate the values. (iii) How this situation can facilitate or hinder the planned change, and to gain Awareness, Commitment, Continuity, and Committed Leadership.
1.2 Myths, legends, and stories	The stories people tell, the myths and the characters are contrary to the intended vision. For example, "you can't," "small batches are highly ineffective."	The stories and characters people mention are consistent with the project's core values, including a commitment to continuous improvement, development, prosperity, efficiency, and customer satisfaction.	(i) Gain insight into the perceptions of success held by the people. (ii) Identify the key stakeholders. (iii) Whether they align with the desired future state. (iv) How this situation may help or hinder the intended change and to gain Awareness, Commitment, Continuity, and Committed Leadership.
1.3 Clear understanding of and alignment with the strategic direction and intent	The staff is unable to provide any information regarding the company's strategic objectives.	The company's business and short- and medium-term objectives are clearly defined and understood by all employees, and they are aligned with the company's overall strategy for success.	(i) Degree of clarity in objectives and strategic plans. (ii) Extent of implementation. (iii) The extent to which the project is aligned with the company's business objectives. (iv) Potential alignment of project sponsors. (v) The relationship that people would attribute to the project in relation to the business. (vi) The business context of the project. (vii) How this situation may help or hinder the intended change and to gain Awareness, Commitment, Continuity, and Committed Leadership.

1.4 Clear understanding of and alignment with operational goals and objectives	They cannot identify operational objectives.	They identify operational objectives that (i) cover all stakeholders, (ii) are interrelated and consistent, and (iii) are consistent with the strategic direction.	(i) Degree of difficulty in adopting the project objectives (ii) How this situation may help or hinder the intended change and to gain Awareness, Commitment, Continuity, and Committed Leadership.
1.5 Open-door relationship	Operational levels have no access to higher levels beyond their boss. The doors are closed, both figuratively and physically.	Hierarchical levels run through the operation and interact with everyone. The doors are open to all, metaphorically and physically.	(i) Degree of difficulty in relating and communicating between levels. (ii) How this situation may help or hinder the intended change and to gain Awareness, Commitment, Continuity, and Committed Leadership.
2. Decision-making process			
2.1 Highly participative and empowering decision-making process. Decentralized and fast decisions.	"No one makes decisions." If they are made, they are centralized in one or a few people, usually at a high hierarchical level.	Decisions are made at the lowest level, taking into account the opinions of all stakeholders. They are supported by data.	(i) Degree of empowerment and decentralization of decision-making, and keys to increasing it. (ii) How this situation may help or hinder the intended change and to gain Awareness, Commitment, Continuity, and Committed Leadership.
2.2 The decision-making process is autonomous and agile	No one makes decisions, or they are made, but "after 5 minutes they are changed."	Decisions are made rapidly and efficiently, with all relevant parties sharing information promptly.	(i) Speed of decision-making and keys to accelerate it. (ii) Degree of agility in making decisions related to the project. (iii) How this situation may help or hinder the intended change and to gain Awareness, Commitment, Continuity, and Committed Leadership.
3. Autonomy, collaboration, and engagement			
3.1 Clear and accepted roles and responsibilities	Unclear roles and responsibilities. Most are maintained without being rigorously enforced. What someone avoids doing is not done at all.	Roles are clearly defined, distributed, and assumed. When "new and unforeseen" situations arise, they are resolved without conflict. When faced with overloads, assistance is requested and provided.	(i) A clear understanding of one's responsibilities. (ii) Awareness of potential conflicts that may arise as a result of the project. (iii) Identifying new stakeholders or the need for new sponsors. (iv) How this situation may help or hinder the intended change and to gain Commitment and Continuity.

(continued)

TABLE 16.1 (Continued)
The table displays the variables for each dimension that can be used to assess the current culture. Also shows the detail of what should be understand to assess, and how to score each of the variables

Variable	Scoring with 1	Scoring with 5	What we want to understand; what it is useful for
3.2 Daily autonomy, initiative, and independence (integrated to 5.6)	People do not feel that they can take the initiative and have autonomy for any kind of task or decision.	Individuals are given the autonomy to test, modify, propose, and make decisions. When authorization is required, the request is promptly addressed with an explanation.	(i) Understand the level of autonomy by area and level. (ii) Understand potential conflicts with project-related responsibilities. (iii) Identifying new stakeholders or the need for new sponsors. (iv) How this situation may help or hinder the intended change and to gain Awareness, Commitment, Continuity, and Committed Leadership.
3.3 Transparent communication	Information, decisions, and the relationship between people and areas are managed outside the scope of "official" channels.	All relationships and communication channels are transparent and clear for all personnel within the organization. They must be conducted in an "official" manner.	(i) The impact of informal networks on the performance, motivation, and morale of individuals or stakeholders. (ii) Who will the staff members turn to for input when evaluating the project. (iii) How the presence of informal networks influences the desired change, including both advantages and disadvantages. (iv) How this situation may help or hinder the intended change and to gain Awareness, Commitment, Continuity, and Committed Leadership.
3.4 Trust and coordination between people and sectors	Silos, islands. Very few people trust others. Areas distrust each other. Communications are written to "preserve evidence." Meetings are not held in a transparent manner, and conflicts are not openly discussed. As a result, conversations often take place in informal settings, such as corridors.	There is a complete willingness to share information and opinions of all kinds. There is a high level of trust between individuals from the same sector and those from other sectors. There is a culture of open and honest communication, where opinions are expressed without fear of repercussions. Issues that may cause discomfort are discussed.	(i) Degree of response to the creation of cross-area working groups. (ii) Potential conflicts between areas and between groups. (iii) Potential limitations on the scope of sponsors. (iv) Degree of trust in potential sponsors of the project. (v) What is necessary to create the sponsors' coalition. (vi) How this situation may help or hinder the intended change and to gain Commitment and Continuity.

3.5 Sense of belonging and pride	People work just to get paid.	There is capacity to reach agreements between sectors and individuals. There are no individuals that can be considered "guilty parties." People are proud to work in the organization. This is true regardless of the level and sector.	(i) The level of difficulty or support that the level of belonging might pose for the project. (ii) Critical factors to work on to improve the sense of belonging. (iii) How this situation may help or hinder the intended change and to gain Awareness, Commitment, and Continuity.
4. Focus on customer			
4.1 Understanding customer needs and competitive position	There is no evidence of an understanding of the internal or external customer. The specific needs of the customer are not identified, and the information provided is vague.	The company has identified its internal and external customers. To gain a clear understanding of their needs, a variety of methods, including surveys, meetings, interviews, and team work are used.	(i) The client's role in the day-to-day operations. (ii) The client's role in each area/level. (iii) The extent to which there may be challenges in fully engaging with clients on their issues, challenges, and needs. (iv) How this situation may help or hinder the intended change and to gain Awareness, Commitment, and Continuity.

(continued)

TABLE 16.1 (Continued)
The table displays the variables for each dimension that can be used to assess the current culture. Also shows the detail of what should be understand to assess, and how to score each of the variables

Variable	Scoring with 1	Scoring with 5	What we want to understand; what it is useful for
4.2 Commitment to meeting customer experience requirements	They fail to provide customer satisfaction data or to acknowledge its existence.	There is a tangible sense of dedication to ensuring customer satisfaction. They provide performance indicators to understand customer satisfaction and key customer experience, both for internal and external customers.	(i) Degree of potential resistance, or support, to respond to customer needs. (ii) The degree of difficulty or support needed to achieve customer orientation. (iii) How this situation may help or hinder the intended change and to gain Awareness, Commitment, and Continuity.
4.3 Commitment to respond to inconveniences	No validation of compliance. No systematic approach to addressing customer complaints or requests. The potential response is often perceived as a nuisance.	A methodical approach to addressing interactions and initiatives from external and internal customers. It is the "natural" and "obvious" expected next step to respond.	(i) Degree of potential resistance, or support, to respond to customer needs, requests, complaints, grievances, and experiences. (ii) How this situation may help or hinder the intended change and to gain Awareness, Commitment, and Continuity.

5. Improvement and Change

5.1 Comprehensive and systematic evaluation of the results, policies, processes, and procedures	There is no indication that any revised procedures, processes, or policies have been implemented. They are written solely to ensure compliance with the relevant standards.	It is standard practice in all sectors and at all levels to review policies, processes, procedures, and results.	(i) The degree of difficulty or level of support required to modify procedures, processes, and policies as needed for the project. (ii) How this situation may help or hinder the intended change and to gain Awareness, Commitment, and Continuity.

5.2 Methodological search for improvements	There is no evidence that any changes or improvements have been made to the procedures, processes, or products.	The improvement of processes, products, and services is based on a systematic and methodological approach to research and development. Indicators and comparisons are used. There is high degree of participation.	(i) Degree of potential resistance or support to the improvement or exploitation of opportunities. (ii) Degree of potential resistance, or support, to the implementation of a methodology to improve or capitalize on opportunities. (iii) How this situation may help or hinder the intended change and to gain Awareness, Commitment, and Continuity.
5.3 Previous changes being successful	There have been no prior modifications, or there are examples of previous modifications that were intended but did not achieve the desired outcome. Frustration.	They provide an overview of previous changes of significance and their outcomes. They also discuss unsuccessful changes, learn and understand a rationale for future decisions.	(i) The keys to a successful or unsuccessful project. (ii) The perception of the individuals involved (pride, frustration, denial, etc.). (iii) How they would react to this new change. (iv) How this situation may help or hinder the intended change and to gain Awareness, Commitment, and Continuity.
5.4 Coordination of projects currently in development or in the planning stages with the day-to-day operations	There is currently no analysis in place that evaluates the relationship, integration, and coordination between projects and with day-to-day operations.	At the executive level, a comprehensive analysis is conducted to integrate the purposes, objectives, scopes, and resources assigned to the various projects. As a result, any conflicts or discrepancies in priorities are resolved. The same process occurs between projects and with day-to-day operation. As a result, priorities for leadership levels are clearly defined.	(i) Potential conflict of interest and competition for resources between the project and with others. (ii) Potential conflicts between the project and day-to-day operations due to the need for dedicated staff attention. (iii) Potential disadvantages of allocating resources to the project. (iv) How this situation may help or hinder the intended change and to gain Organizational Leadership, Awareness, Commitment, Knowledge, and Continuity.

(continued)

TABLE 16.1 (Continued)
The table displays the variables for each dimension that can be used to assess the current culture. Also shows the detail of what should be understand to assess, and how to score each of the variables

Variable	Scoring with 1	Scoring with 5	What we want to understand; what it is useful for
5.5 Changes are sponsored by individuals in senior management	Sponsors of current or past projects are not identified.	Project sponsors are clearly identified by the people. They have the political standing and influence for the role.	(i) An awareness of the role and importance of a sponsor in a significant project. (ii) The willingness to form a sponsors' coalition. (iii) How this situation may help or hinder the intended change and to gain Awareness, Commitment, Continuity, and Committed Leadership.
5.6 Risk analysis	No risk assessment has been conducted.	In the event of significant changes, a comprehensive risk analysis is conducted. In the case of minor changes, the relevant parties provide an opinion-based analysis of potential risks.	(i) The level of confidence that the project's major risks will be anticipated. (ii) How this situation may help or hinder the project. (iii) How this situation may help or hinder the intended change and to gain Commitment, Knowledge, Continuity, and Committed Leadership.
5.7 Error tolerance	The phrase "you can't go wrong" is a common one. People tend to avoid taking risks. When mistakes are repeated, there is often no action taken to avoid them.	Mistakes are used as a basis for drawing conclusions and implementing measures to avoid similar errors in the future. Root cause analysis is conducted. These measures can be corrective or preventive, and they can be taken at the individual or group level. The error is viewed as an opportunity for learning and improvement.	(i) The degree of tolerance for errors. (ii) The concept of error, ignorance, penalization, and learning. (iii) How this situation may help or hinder the intended change and to gain Awareness, Commitment, Knowledge, and Continuity.

5.8 Offices, plants, and commercial areas "speak"	No information is available, and it is difficult to find one's way around the rooms (offices, floors, etc.).	There is no way to get lost or not understand the spaces; everything is clearly identified: security aspects, areas, management. Key information about stakeholders and the operation (office and plant) and the company is visible.	(i) The consistency of the physical spaces with the values and objectives. (ii) How the project can help or hinder communication, teamwork, agility, improvement, and quality of work life.

6. Technology and innovation

6.1 Continuous technology improvement with a focus on results	Systems are not integrated and not "leveraged"—e.g., partial use of CRM or ERP. The data and information derived from these systems are suspect. The use of alternative systems—such as Excel—is widespread. There is no measurement of the performance status of the systems.	They are thought and decided (i) from the need and the solution, not from the dazzle of the technology itself; (ii) based on the needs of the stakeholders. Use updated and centralized systems to maintain and use information and project scenarios. The reliability of the information is ensured. Information is shared with suppliers and customers. The level of use and performance of the systems is measured and improved.	(i) Degree of ease or resistance to systematize activities and processes. (ii) Degree of ease or resistance to incorporating new technologies into the project. (iii) How this situation can help or hinder the intended change, and to gain Commitment and Continuity.
6.2 Creativity and innovation	When changes occur, they are based on known solutions with no intention of being creative.	Practices are used to generate new and creative solutions to challenges. Techniques such as Design Thinking, Blue Ocean Strategy, Triz, Gamification, AI, and engaging with industry leaders are employed to foster innovation. There is a commitment to "embracing new ideas."	(i) The conditions for implementing new approaches, methods, and techniques that result in innovative solutions—at least within the organization. (ii) How this situation can help or hinder the intended change, and to gain Commitment and Continuity.

(continued)

TABLE 16.1 (Continued)
The table displays the variables for each dimension that can be used to assess the current culture. Also shows the detail of what should be understand to assess, and how to score each of the variables

Variable	Scoring with 1	Scoring with 5	What we want to understand; what it is useful for
7. Agility			
7.1 The projects, particularly this one, are part of a strategic plan, are supported by a robust methodology, and are agile.	Agility is not under consideration.	The project is aligned with the strategic objectives in a clear and defined manner. Stakeholders' needs and expectations are under consideration. The selected methodology prioritizes agility and effectiveness.	(i) How this situation can help or hinder the intended change, and to gain Commitment and Continuity.
7.2 The encouragement of proactivity and freedom to experiment under one method is a key objective	There is no methodology in place to analyze, test, and accelerate changes.	A systematic approach to analyzing, testing, accelerating, and measuring the impact of implemented changes is currently in use. There is a preference for agility and methodical learning.	(i) The degree of acceptance of the methodology to be used. (ii) The degree of agility of the organization. (iii) The speed at which the project could be deployed. (iv) Areas and levels to be supported or with greater difficulty. (v) How this situation can help or hinder the intended change, and to gain Awareness, Commitment, Knowledge, and Continuity.
8. Reinforcing behaviors			
8.1 Systems of formal and informal recognition	No positive recognition. For some people, "negative" recognition.	Formal recognition system consistent with the project's values and criteria.	(i) How this situation can help or hinder the intended change, and to gain Commitment, Knowledge, Continuity, and reinforcement of behaviors.
8.2 Promotion system in line with values and guidelines	People are promoted for cronyism, convenience, or without known criteria.	People who are recognized as having merit consistent with the values and goals of the project will be promoted.	(i) Available resources and system. (ii) Consistency of HR systems with the project. (iii) How this situation can help or hinder the intended change, and to gain Commitment, Knowledge, Continuity, and reinforcement of behaviors.

8.3 Evaluation system aligned with values and policies	A systematic evaluation that aligns with the project's values and objectives is in place.	There is no formal evaluation process in place. When one is introduced, it is often questioned for lack of objectivity.	(i) Available resources and system. (ii) Consistency of HR systems with the project. (iii) How this situation can help or hinder the intended change, and to gain Commitment, Knowledge, Continuity and reinforcement of behaviors.
8.4 Training system aligned with values and policies	Training system focused on improving performance and future competencies. Covers operational, relationship, and leadership aspects. Is aligned with strategic and operational values and objectives. Employees and their managers are responsible for it.	There is currently no training system in place. When training courses are initiated, they are usually initiated by the HR department.	(i) Habit and appreciation of training. (ii) Available resources. (iii) Consistency of HR systems with the project. (iv) How this situation can help or hinder the intended change, and how to acquire the knowledge and skills needed for the project—for both the transition and final stages of the project.
8.5 Physical spaces designed for comfort and sharing	The space encourages information sharing, communication, and teamwork. There are meeting places and communication spaces that everyone uses. The atmosphere is pleasant. Common areas are well maintained. Floors, offices, restrooms, locker rooms, and common areas are equally well maintained.	The physical spaces are not conducive to information sharing, communication, or teamwork. The entire space is divided by doors and partitions.	(i) Alignment of physical spaces with values and goals. (ii) How the project can help or hinder communication, teamwork, agility, and quality of work life.

Note: The dimensions used may not have the same weight for different types of projects. Depending on the type of project, the survey method may change. For example, the level of involvement of all employees in the development of a project to achieve an Agile or Lean culture must be higher than in a project to upgrade to SAP Hana.

ADDITIONAL COMMENTS

- The objective of the survey and meetings is not to transcribe the statements made by the interviewees. The conclusions will be based on the interpretation of the obtained data. It is not a compilation that takes what people say as true but a search for data to get more information. It is not a matter of repeating what people say.
- Possible questions from executives, managers, bosses, and even sponsors—such as "Who said that?"—should not be answered. This should be clarified prior to interviews or surveys.

WHO IS RESPONSIBLE FOR THIS?

I observe that the HR team in collaboration with the support facilitator team is responsible for this. It is important to note that we are not discussing two distinct projects. Rather, we are examining the same project from three different perspectives: technical, change, and project. The culture survey basically requires competencies related to the social field—usually in the HR areas—but those with a different perspective can raise different questions because of their greater knowledge of the changes that should take place at the operational level.

SURVEYING A LIMITED PROJECT

In the event that time or the expected impact on people are limited, an alternative approach is to develop a workshop with input from all relevant areas. This can be done by following the questions in Table 16.1 and incorporating the conclusions into a Stakeholder Analysis spreadsheet (see Figure 19.3).

Also, in projects with a limited number of participants or minimal scope for change, a survey, even a sample survey, with questions based on those in Table 16.1, may be sufficient to supplement existing knowledge about the culture.

> *The culture survey typically requires competencies related to social issues, typically within the areas of HR.*

CHARACTERIZATION OF THE CULTURE

The characterization process is ambitious because it involves synthesizing or typifying elements that are, by nature, exclusive. However, it is a crucial step in developing a comprehensive change plan. I have already shared the key dimensions. Characterizing the culture means understanding and synthesizing the primary factors that differentiate it in relation to the project.

> *Characterizing, now, the culture means understanding and synthesizing the primary factors that makes it vulnerable it in relation to the project.*

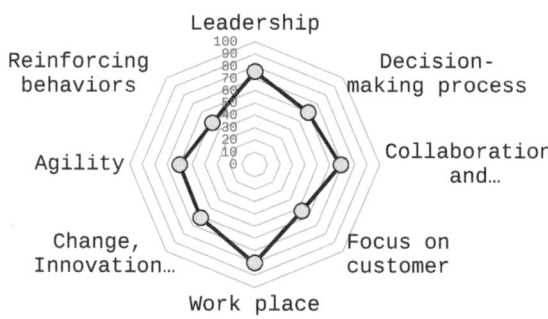

Cultural assessment

FIGURE 16.1   Chart with the results for each of the eight dimensions of a culture survey. Leadership and Workplace are those with the higher scores, and the Decision-Making process, Focus on Customers, Agility, and, mainly, Reinforcing Behaviors are those with the lower scores and to be considered for how they could impact the project and for improvement. The dimensions used were customized by the organization for this specific project.

For example, in a Six Sigma project, whether mistakes are used as opportunities for improvement or punished, whether highlighting problems is valued or hidden, or whether data is used to make decisions or ignored.

A final suggestion: focus on evidence that is relevant to the transformation project. It is not about changing the culture completely or for the sake of changing it; it is about looking for characteristics that will enable the improvement or transformation project to move forward.

> *It is not about changing the culture completely or for the sake of changing it; it is about looking for characteristics that will enable the improvement or transformation project to move forward.*

TYPIFYING CULTURE

In addition, it is interesting and especially clarifying for those who have to interpret the culture to classify it, for example, as collaborative, hierarchical, controlling, competitive, flexible, or proactive. Karl Jung's personality archetypes are extremely useful in this regard.

IT IS NOT SIMPLY A MATTER OF ORGANIZATION; IT IS ALSO A MATTER OF PEOPLE

Cultural change is not a cost-free process; it entails costs. Many of these costs are typically social in nature. It is important to recognize that when we discuss changing culture, we are referring to modifying the beliefs and behaviors of individuals who have been selected to behave in the manner they do. People have the right to think and say, "I do not want to change that because this is what your hired me to do."

> *We are talking about changing the beliefs and behaviors of people who have been selected to behave as they do and to believe in what we are now trying to change.*

From our perspective, corporate social responsibility is demonstrated in the choices we make, in what we do, and in how we interact with those who are not part of the change.

ABOUT THE TABLE 16.1

In order to provide an approximation, the values assigned to the responses may vary from 1 (representing the most difficult situation for the project) to 5 (representing the most favorable situation). The average of the values assigned to each component may be used as the value assigned to the criterion. These values are then used in the overall assessment and lead to the overall project strategy and plan—see Chapter 18 "What Could You Be Facing."

Note: Technically, the values should not correspond as if it were a continuous scale, because it is not. Nevertheless, we use it like that to simplify the understanding. 0.1 will replace 1, 0.25 will replace 3, and 1 will replace 5 in a scale used by QFD.

NOTES

1 John P. Kotter, *Leading Change*. Harvard Business Review Press, 2012.
2 Cheryl M. Jekiel, *Lean Human Resources. Redesigning HR Processes for a Culture of Continuous Improvement*. CRC Press, 2011.
3 Edward De Bono, *Six Thinking Hats*. Back Bay Books, 1985.
4 Luc Mayrand, *The Imagineering Way*. Disney Editions, 2005.
5 Peter Senge, *The Fifth Discipline in Practice*. Granica, 1995.
6 Edgar H. Schein, *Culture and Leadership*. Jossey-Bass, 2010.
7 Edgar H. Schein, *Culture and Leadership*, Jossey-Bass, 2010.
8 Simon Clatworthy, *The Experience-Centric Organization, How to Win Through Customer Experience*. O'Reilly, 2019.
9 Luciana Paulise, *We Culture, 12 Skills for Growing Teams in the Future of Work*. Quality Press, 2022.

17 Readiness to Change

Raúl Molteni

WHAT TO ASSESS CHANGE READINESS FOR

It is essential to start the project with basic and adequate preparation to evade negative consequences caused by risks, mistakes, and resistance that could have been avoided. Then it is crucial to structure a core competency development plan to be executed before kickoff.

GLOBAL READINESS FOR CHANGE

Once the project has been fully defined and the relevant background information has been gathered, we must assess the organization's ability to successfully undertake the project. The project may be complex, but with proper preparation, it can be managed effectively. For instance, the plan may have been thoroughly analyzed for potential risks, the sponsors may have received training, the organization may have a history of successfully implementing changes, and the staff may have agreed to and strongly supported the process.

On the contrary, there may be a less complex change that has not yet been clearly defined and is being promoted only by a new general manager. In addition, the organization may have failed in several previous attempts at implementing changes and may lack a structured plan. Both situations require different levels of preparation and project plans.

The level of preparation will ultimately determine the level of support needed for the project before it is launched—or contingency measures if it has already been launched—and, therefore, the level of activity required before the kickoff.

The level of preparation is the one that allows you to define the support and activities before the launch.

DOI: 10.1201/9781003544807-22

UNDERSTANDING INITIAL READINESS

This is a process of data collection and analysis conducted in accordance with preestablished criteria.

Our proposal is based on three fundamental criteria: (i) real need and Organizational Commitment; (ii) degree of structuring and completeness of the plans; and (iii) background. The questions described in Table 17.1 serve as a guide for the evaluation process. It shows the questions asked to assess the level of prior preparation that the organization has. It leads to the identification of aspects that should lead to actions before the kickoff.

TABLE 17.1

Questions asked to assess the level of prior preparation that the organization has, facilitating identification of aspects that should lead to actions before the kickoff

Feature	No/Not so	Completely and rigorously	Score
Real need and organizational commitment			
There is a real, shared sense of need and urgency among owners, directors, and executives.			
There is a designated "volunteer" sponsor at a high hierarchical level who is accepted by all to lead the project.			
The project is highly aligned with the organization's future plans and positioning.			
Executives accept their responsibility, understand their role, and have initiated their own preparation and training to lead the project.			
Managers and supervisors share the need to improve the project. There is a conviction that things need to work differently in the near future.			
Degree of structuring and completeness of the plans			
The project has objectives and planning, at least for its first stages.			
All leaders—with technical, social, and project— have the appropriate training and follow common goals.			
The goals, principles, and techniques of the projects are properly understood.			
As part of the planning, the resources that will be needed for the project have been considered and allocated, at least for the early stages.			
As part of the planning, risks have been considered— aspects that could cause the project to "fail."			

TABLE 17.1 (Continued)
Questions asked to assess the level of prior preparation that the organization has, facilitating identification of aspects that should lead to actions before the kickoff

Feature	No/Not so	Completely and rigorously	Score
Background			
The changes implemented in the last five years have not resulted in high costs for people.			
The change efforts are coordinated with other projects that are under development or about to start.			
One of the factors that scores well in the climate survey is the one related to employee autonomy and initiative.			
The survey does not clearly show the existence of "silos." Interdepartmental teamwork is appreciated.			
Overall result			

Each question can be rated on a scale of 1–5, with each rating corresponding to the descriptions provided. The score values for each question are totaled to provide a score for each criterion.

A total score below 50% indicates that the project requires strengthening and restructuring before launching. Without these improvements, the project is likely to face significant challenges in execution time, achievement of objectives, and cost management. Compliance percentages below 70% indicate a moderate risk and the need to address the lowest scoring axis or axes in greater depth. Higher values indicate a low-risk project (see Figure 17.1).

TYPE OF CONCLUSIONS

Conclusions should not be used to present a picture of the situation or to complain about it. They should be used to provide feedback and improve the deployment plan. These are the kinds of conclusions that analysis and data can lead to:

- The level of project sponsorship is sufficient to initiate Lean implementation and deployment.
- Although the project is medium risk, change management needs to be carefully planned and monitored in light of past and future projects.

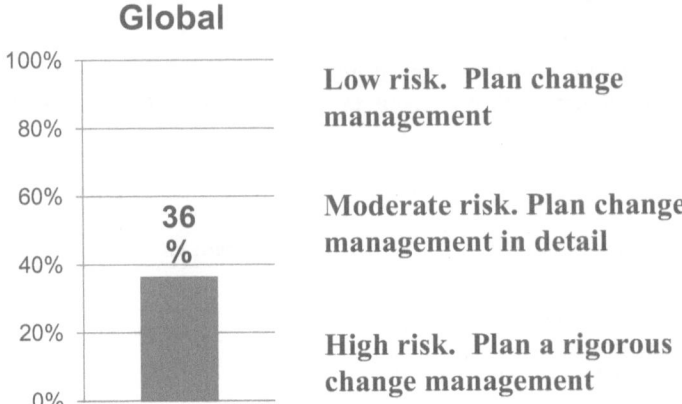

FIGURE 17.1 The graph of readiness value in a service organization—a leader in its industry—and suggestions for actions on the right vertical axis. The score of 36%—shown on the scale at the right—suggests that the project has moderate risk. However, it also suggests planning the change management in detail.

The conclusions are not meant to show the picture of the situation; they should be used to provide feedback and improve the deployment plan.

SUGGESTIONS

I see practices that undervalue this analysis in the desire to move quickly because concerns about prior preparation could be seen as resistance and delay or cancellation of the project. Therefore:

- Analysis should not be seen as a bureaucratic exercise.
- It is good practice to validate data and conclusions with references.
- It is advisable to maintain an objective but empathetic position.
- It is preferable to not overestimate the capacity of the organization to bring about cultural change.

18 What Could You Be Facing

Raúl Molteni

WHY SYNTHESIZE INFORMATION

Synthesizing information is essential to gain clarity on priorities and areas of focus, and as a result, develop strategies and plans that are "tailored" to the situation.

INFORMATION TO BE CONSIDERED

1. About the Culture

What to consider: The main characteristics of the culture that relate to the project and changes to be implemented.

How: The main characteristics of the culture (Chapter 16, "Understanding your Current Culture") accompanied by the main archetype, or archetypes, that have emerged.

2. About the Stakeholders

What to consider: The stakeholders and the impact on them that has resulted from understanding the project, and for each of them, the aspects with the greatest impact, which will be negative and positive for the project (Chapter 15, "Understanding Your Project").

How: The list of stakeholders ordered by degree of impact and with comments on the cultural and operational aspects that will be affected (Table 18.1).

3. On the Level of Preparation

What to consider: Readiness-related aspects with higher positive and negative scores (Chapter 17, "Readiness to Change").

How: The list of the questions—aspects—that have received the lowest rating in the analysis.

4. About Potential Sources of Resistance

What to consider: The people and groups identified, and the type of impact and response expected of them as a product of the change. Understanding that even the promoters of the change will show resistance to maintain complete coherence with what the change promotes.

DOI: 10.1201/9781003544807-23

How: An organizational chart that identifies the function or people with a color related to the type of reaction expected to the change (Figure 18.2). For example, with different shades and from green to red, identify those who are seen as having a profile of:

- *Sellers* of change, who will act as promoters.
- *Supportive*, when needed, without being proactive or promotional.
- *Indifferent*, who will ignore it—some pretending to comply and others simply trying to "escape".
- *Resistors*, who will try not to comply with the demands of change.
- *Opponents*, who will overtly or covertly express themselves with arguments to deny the change.

The facilitator team can simplify the categories or add intermediate categories, such as "Visible Opponent" and "Hidden Opponent," if necessary. It is equally acceptable to use a table with names, the aforementioned profile, and a summary of the distinguishing characteristics of the individual or group. The specific approach ultimately depends on the organization's culture regarding the confidentiality of data and information.

This is not an evaluation of the individual in question, nor is it a rating of their performance or value to the organization. It is an estimation of how the individual will respond to the proposed change. It is also important to consider that this individual was previously incorporated, supported, and likely trained, promoted, and rewarded for maintaining a behavior that is now being questioned.

> *This is not an evaluation of the individual in question, nor is it a rating of their performance or value to the organization.*

As I have said, those who identify themselves as Resistors and Opponents are key to identifying opportunities to improve the vision and plans for change and to understanding the source of the resistance.

> *Resistors and Opponents are key to identifying opportunities to improve the vision and plans for change and to understanding the source of resistance.*

FOCUS OF WORK FOR PLAN DEVELOPMENT

The final step, using Kurt Lewin's Force Field model, is to synthesize the key aspects—beliefs, values, behaviors, cultural characteristics, readiness conditions, and personal responses—that are important to the organization:

- They will help create awareness, gain commitment, and implement change.
- They will hinder change.

Driving Forces

- Personal interest and promotion by the CEO.
- VP with disposition/ attitude.
- Top-down organization facilitates for alignment.
- Background (Central American Unit) with results recognized as very positive by the personnel involved.
- Support from the management of the Central American Unit and consultants.
- Engineers present no resistance to participate.
- Engineers know the process.
- Basic data exists.
- HR support for Agile.

Restraining Forces

- Differences related to the sense of urgency and use of methodology.
- Evidence of misalignment of objectives between HR and the Operations Project.
- Changing priorities.
- VP without knowledge of the methodology or participation from the beginning.
- Culture in the Andean Unit is different from that of Central America.
- Culture is "accustomed to letting the wave pass." "They are not going to convince us". Feeling that improvements are not going to be implemented. "In 2020 there was a similar project and it came to nothing".
- There is no participation from those who are outside the improvement teams.
- Product Owner with many responsibilities -without time-.
- Union with potentially conflictive position.
- Position and attitude of suppliers?

FIGURE 18.1 An analysis of the Driving and Restraining forces for change, as described by Levin's Force Field. The Driving forces result from characteristics of the organization's culture that should be conducive to implementing the proposed changes. The Restraining forces are those characteristics that are expected to impede progress.

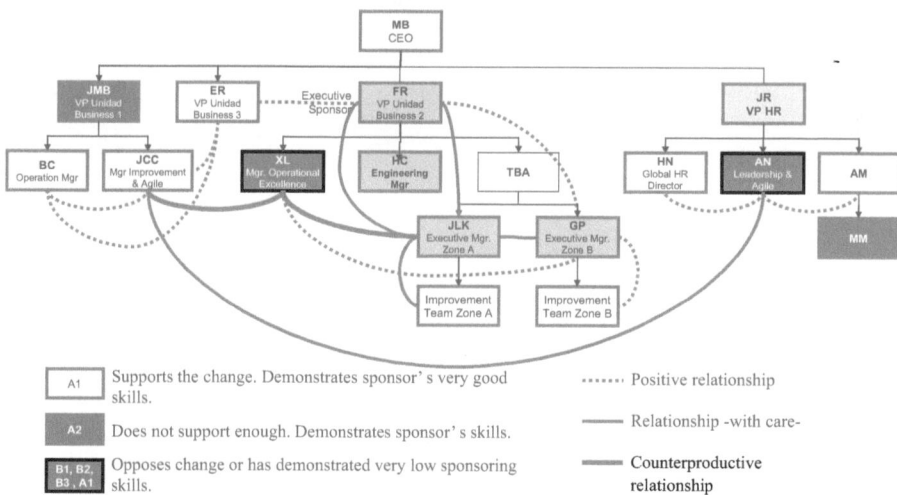

FIGURE 18.2 The individual positioning analysis of the key personnel for the project. The organizational chart shows different colors for each person, each color indicating opposing, supporting, or fully supporting the change. The thickness of the lines connecting the individuals represents the type of relationship between them—including communication, mutual trust, and teamwork.

TABLE 18.1
The criteria to guide the impact that the project and its changes will have on each stakeholder group

Stakeholder	Group size	Importance of its contribution to achieving change (A-M-B)	How does change affect it?	How does it perceive the change and how it will react?	Current level of commitment	Approaches to group engagement (who, when, and how)

The elements synthesized at this stage of the process will help develop the strategy and main elements of the action plans. See Figures 18.1 and 18.2 for more details.

In addition, a risk analysis may identify new considerations for the change plan, such as difficulties in obtaining resources or maintaining a minimum production level while pilot testing a redesigned process. The key objective is to interpret the findings and determine the best course of action. While filling in forms and Excel sheets can be useful, they should not be the primary focus, as they can be costly and time-consuming.

All the information can be efficiently recorded as shown in Table 18.1 and will be used when designing the change management plan.

Part VI

Designing for a New Level of Performance

This section outlines the second phase of the roadmap. The objective of this phase is to develop a strategy and plans based on the findings of the previous section regarding the initial situation.

This section addresses the essential elements and recommendations for the design of the strategy and the main plan, as well as the subordinate plans related to clarity and alignment, communication, education and training, coaching, resistance management, recognition and reinforcement of behaviors, and the indicators to be installed.

The deliverable of this phase is to determine the optimal approach for executing the project. This will include identifying key personnel, delineating responsibilities, and establishing performance indicators to assess progress.

DOI: 10.1201/9781003544807-24

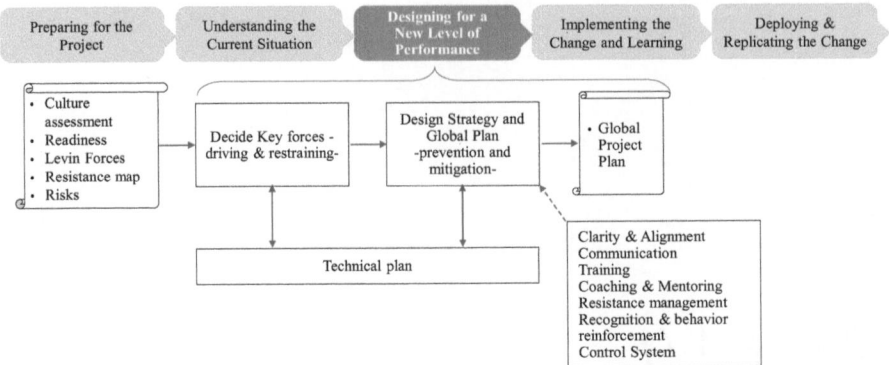

FIGURE 0.6   The two steps of the Designing for a New Level of Performance phase: Deciding the Key Forces to focus on and Designing the Strategy and Global Plan. It also lists the resources to be used. The steps are discussed in Chapter 19, and the resources in Chapters 20 to 27.

19 Designing the Global Plan

Raúl Molteni

The mere fact that something can be done does not necessarily mean that it should be done.

WHAT WE EXPECT TO GET FROM PLANNING

The expected outcomes of planning are:

- *Strategy Map*: helps to "see" the strategy.
- *Gantt or PERT diagram*: shows the activities and their interrelationships. It could be divided into sections corresponding to each function of the organization to make it easier to identify who is responsible, or it could be divided into stages to better see the relationship between functions and silos. It is a "living" document that is constantly updated according to the results of monitoring, evaluations, lessons learned, and unforeseen circumstances that adjust the designed plans.
- *Dashboard* with key project indicators (see Chapter 27, "Metrics and KPIs"): measures the effectiveness of the planned actions and systematically provides feedback to the plans.

POWERFUL PURPOSE

The final goal should be to learn how to change, not just how to manage the change of one project.

ABOUT THE STRATEGY

I call strategy the logic that makes up the few milestones that are central and critical to achieving the project's objectives. They are the pillars on which the plan is built.

DOI: 10.1201/9781003544807-25

> *I call strategy the logic that makes up the few milestones that are central and critical to achieving the project's objectives.*

A biotech company elected to implement SAP HANA. In an informal discussion with the director and HR manager, it was revealed that two previous attempts had not been successful. The managers, along with the IT personnel, were confident that they now had a firm grasp on the situation. Nevertheless, we conducted an accelerated culture survey to gain insight into the power dynamics between managers. The three managers were genuinely convinced that everything was under control. However, another director—Operations director who had the greatest influence over the owner—was convinced that the installation of SAP was not the organization's primary focus. All three parties—the HR director, HR manager, and IT manager—then acknowledged that the involvement of this other director had contributed to the two previous failures. A director and two managers were initiating the implementation of an IT system without adequately assessing the influence of the individual who had previously halted the project on two occasions. The rationale behind this decision was the perceived "logical and rational" advantages of the project, namely the installation of SAP. While a culture survey may not have been a prerequisite for identifying resistance, it undoubtedly provided insights into the underlying dynamics.

The subsequent phase was to ascertain the probability of proceeding and the optimal methodology for doing so. The analysis of a relationship map revealed a "network of influences" that could be leveraged to secure the support of the operational director. The system was successfully installed within the agreed timeframe and budget. The change management plan yielded numerous effective actions that helped maintain control over the agenda, time, and costs. Examples included coaching process owners and advancing and adapting the training of key users. The question remains, however, whether the design of the implementation plan itself was more important, or whether it was the steps taken to reach an open conversation with the Operational director that proved to be the key factor. Once the Operational director demonstrated awareness and commitment, they were able to rapidly align over 60% of the personnel involved in the project.

When planning a project, several questions usually arise. Some of the answers could become the pillars of the strategy.

- What should be the project's main objectives and metrics of success?
- When, who, and how will the board, management, and the CEO be aware that the project depends on them changing this or that behavior?
- Who are the key personnel that everyone will pay attention to when giving opinions on the project?
- What makes the sponsors' coalition strong?
- What will the simpler, rational, emotionally motivating, and aligned interim opportunities and quick wins be?
- Who and where are the early adopters?
- Where is the likelihood of quick wins greater?

- Will there be fireworks or a low-profile launch of the project?
- How will HR and Operations be integrated if they are working separately?
- What will be the position toward the Union?
- How will the consultants already working with the company be integrated?

Answering these questions is easy; framed in the project, they require meditation.

SUGGESTIONS FOR CREATING THE STRATEGY

- Do not rush. Take time to think about the strategy. Do not think of it as something bureaucratic. These are the keys to success.
- Think at a high level. Milestones are more important than the plans themselves.
- Involve the stakeholders. Listen to them, analyze their points, and review them with their participation.
- Maintain an objective but empathetic position.

ABOUT THE MAIN FOCUS

The first step is to gather all the information worked on until now. That will include the Culture Assessment, Readiness to Change, Resistance Map, and the Risk Analysis—which comes from the technical analysis.

In addition, the main forces from the Levin analysis—Driving and Restraining Forces—should be decided based on their impact on the project.

ABOUT THE PLAN

Milestones should guide the basic structuring of the plans. Milestones mark the objectives and should be the primary focus rather than the plans themselves. It is not sufficient to complete 100% of the planned actions if they do not facilitate progress toward the milestones and the final objective.

> *It is inadequate to fulfill 100% of the planned actions if they do not facilitate progress toward meeting the milestones and objectives.*

The actions that make up the plan must follow logic and not because it is "what everyone does." For example, it is well known that communication must be continuous and that we must consider the purpose, recipients, timing, and method of communication. I agree. However, there is an additional factor to consider: the reason for communicating. Less emphasis should be placed on "what to communicate" than on "what to communicate for."

> *More emphasis should be placed on "what to communicate for."*

It is acknowledged that training will need to be provided to all relevant parties. Once more, the question "What for?" is the key that determines what, to whom, when, and how.

Questions that help visualize actions are as follows:

- How do we approach each function, each stakeholder, and each referent?
- What are the key communication points and timing?
- What recognition should there be, to whom, when, for what reasons, and what is the process for deciding and delivering it?
- How should future promotions be handled? Who should be promoted and why?
- Will there be a kickoff?
- How will sponsors be prepared?
- Who will do the coaching, how, and for what purpose?
- Who, how, and to what end will individual resistance identify? And group resistance? And organizational resistance?
- Will the project have its own identity, name, logo, and artifacts?

ABOUT THE METHODOLOGICAL RESOURCES

Communication and training are important in change management. However, they are not the only factors. Individuals typically go through several stages when confronted with a change: gaining Awareness of the need for change, accepting the Commitment to actively participate, acquiring the necessary Knowledge and Skills, maintaining Continuity of participation, and perceiving the Leadership and support of the organization.

- *Gaining Awareness* is highly dependent on the information and rationale behind the vision and proposal for change. Communication—not just information—is key, as is who communicates it and how. For example, if the change has an impact on people, the fact that it is communicated by a supervisor in the midst of the silence of senior management may mean that the person proposing the change is not ensuring continuity or integral support. Therefore, it pre-announces a limited life for the change.
- *Gaining Commitment* brings us to the person. When a person is asked to behave differently—for example, to spend more time in meetings, to make decisions when he/she used to ask for approval, or to question operational aspects that he/she used to accept without question—the focus of attention changes. Awareness has to do with agreeing or disagreeing with the change, and Commitment has to do with getting involved, risking mistakes, and exposing oneself. The great referent is no longer the CEO but the boss. Communication is still a fundamental resource, but now it is with your boss. It may also be necessary to help him or her reflect to address the resistance and questions that arise; coaching and mentoring come to the rescue, as do organizational alignment activities.

- *Acquiring Knowledge and Competencies* requires learning and unlearning. Now, it is necessary to have segmented training activities which are designed according to the real needs of the project. And with concrete application of the new Knowledge or Skills to the tasks and processes in which they intervene. It may also be necessary to provide assistance when there are particular difficulties. Coaching and Mentoring, as well as some individual training, may be helpful.
- *Demonstrating Consistency and Continuity* in applying the new behaviors is imperative. Here communication and training will continue but with less emphasis. Understanding inhibitions and motivations is the most important factor in achieving continuity. For example, it is essential to identify the reasons that prevent the person from understanding more deeply what and how to do in the future and how to stop doing what he or she is currently doing. It is important to find the details that prevent the standardization of new methods and the simplification of tasks, processes, and procedures. It is mandatory to eliminate the risk of incidents and accidents. At this stage, it is essential to support middle management because they are the ones who most condition the adoption—or rejection—of new behaviors.
- *Leadership and Organizational Commitment* must be maintained throughout all phases. Their visibility, commitment, and control of the project and change process may require coaching and mentoring, training, alignment, and a great deal of effort. As part of this, the sponsors' coalition must be maintained, as well as the message that the project belongs to the organization and not to any one individual.

Going through these several stages when confronted with a change require much more than communication and training. Several other resources will be needed; see Chapters 20 to 27 for more information (see Figure 19.1).

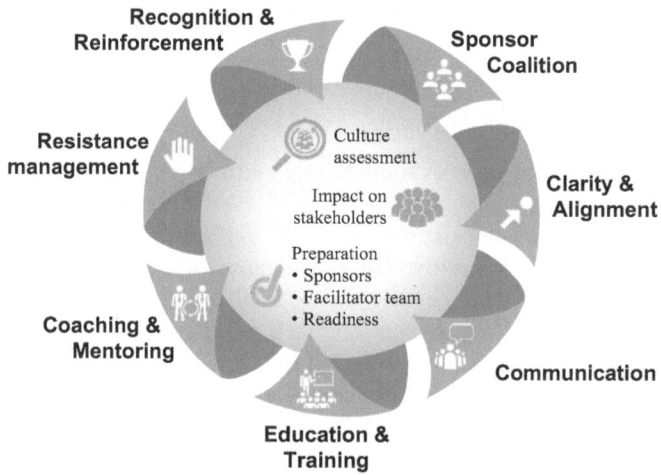

FIGURE 19.1 The resources available to help people achieve Awareness, Individual Commitment, Knowledge and Competencies, Continuity and Routines, and Leadership and Organizational Commitment—five stages of change maturity.

ON FINANCIAL AND TIME RESOURCES

The project is likely to have an estimated ROI or economic benefit calculation based on the opportunity gap to be captured. It is also possible that the organization's culture has not led to such an estimate. However, it is crucial to ensure that indispensable resources are not taken away from the project, particularly in two key areas:

*   Staff time allocated to training, design, coordination, and execution of planned actions.
*   The investment required to sustain the project.

DESIGNING THE GLOBAL PLAN

Global Plan refers to a single plan that emerges from weaving together the activities necessary to move forward:

*   The "technical" project. The one that significantly impacts what people believe, think, and do. Whatever the intention is—how to make the organization agile, certify ISO 9001 to ensure quality and effectiveness, develop improvement projects using Six Sigma or Lean, or drive a digital transformation.
*   Change management. To ensure that as many people as possible adopt new behaviors and create new habits in the shortest possible time.
*   Managing the project itself. For example, following PMI guidelines and steps.
*   Managing other projects. Ongoing or to be started, such as the integration of the digitalization of production equipment.

> *It's not about unifying Gantt charts; it's about ensuring a single project. It is not a project of each function; it is a project of the organization.*

INTEGRATE AND INTERTWINE ACTIVITIES

Intertwining activities involves examining the interrelationship and potential synergy between activities from different perspectives. Should a manager be asked to provide support to improvement teams in his area, or should he be shown the results achieved in another area first? Is it positive, in this organization, to mention what "Toyota does it this way," or will it be seen as "we are not from the automotive industry" and have a negative impact? What obstacles might arise from implementing Kaizen, both in the groups that currently utilize it and in other potential future sectors, and what should be changed to overcome these obstacles? What is the anticipated response from managers when they are assigned the role of Process Owner for an Agile cell so the training can be adapted? Who are the relevant stakeholders in a sector to analyze with them the steps to develop a Six Sigma project?

A Touch of Creativity

Do not copy others. You must find "caterpillars."[1] The facilitator team and sponsors' coalition should not call any proposed ideas ridiculous. They should build on them to come up with better ones. Use "What if ... " instead of "No, because ... ."

> *"The day before something is a breakthrough, it is a crazy idea."*[2]

Listen, Talk, and Co-Create

This is not the plan of a group of facilitators, a sponsor, a sponsors' coalition, or the owner. Many more individuals will be involved. It is essential to abandon the belief that one possesses absolute "truth" or knows what is "right." When we acknowledge the existence of multiple interpretations, we seek the most valuable for the purpose, which allows us to identify new possibilities and realities. What was previously considered impossible may now be seen as probable. This is the shared vision and shared project.

ON THE GENERATION OF THE GLOBAL PLAN (SEE FIGURE 19.2)

1. Consider all conclusions by now as a checklist and guide:
   - The driving and restraining factors of cultural analysis.
   - Opportunities identified through the readiness analysis.
   - The expected reactions of key stakeholders and Referents with the greatest impact.
   - Map of individual reactions.
2. Design actions to initiate and reinforce change Awareness (see Chapter 20, "Checklist for the Planning," for guidance).
3. Continue the plan by elaborating on actions to build Commitment, Knowledge and Competencies, and Continuity and Routines (see Chapter 20, "Checklist for the Planning").
4. Install maturity indicators for key stakeholders and groups.
5. Plan actions, from the ground up, to maintain and make visible the Leadership and Organizational Commitment.

Information to Be Used

Chapter 15, "Understanding Your Project," Chapter 16, "Understanding Your Current Culture," and Chapter 18, "What Could You Be Facing," provide the necessary details to develop the strategy and plans. These chapters also serve as a guide for creating the plan discussed in Chapter 20, "Checklist for the Planning."

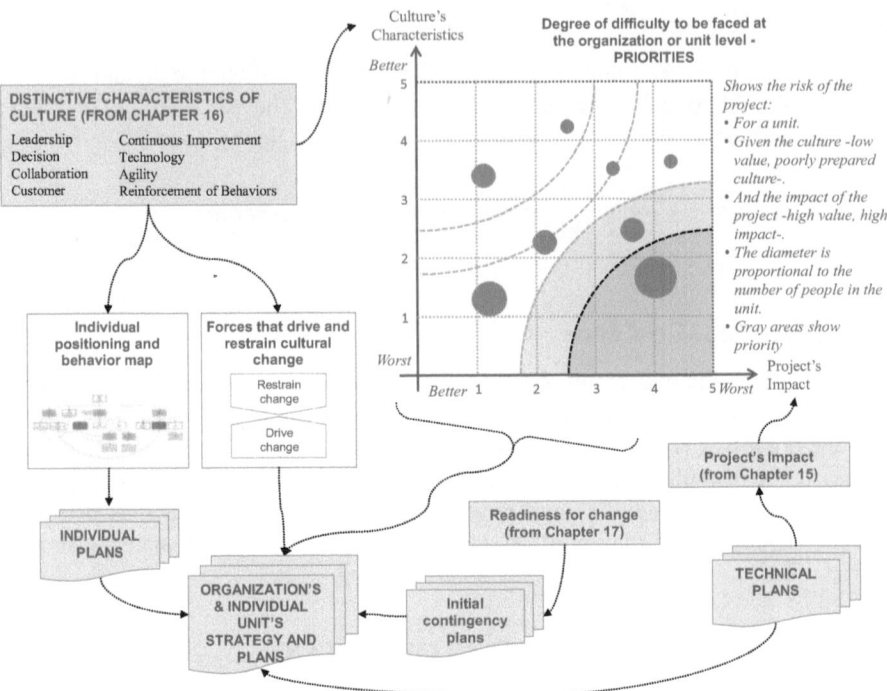

FIGURE 19.2 The integration and sequence of information and deliverables used to develop the strategy, as well as individual, unit, and organizational plans. It can be seen that cultural and project impact—understood from the analysis of the technical plans—allows for the identification of priorities. Those priorities, individual positioning, driving, and restraining forces combined will lead to the strategy and individual and group plans.

Stakeholder	Role in project	Change impact on stakeholder	Mind set	Expected resistance or reaction	Required Support to project	Strategy to use	Responsible	Date	How and when to look for changes

References

Role in project   1: Spectator. 2: Indirect influencer. 3: Direct influencer. 4: Makes key decisions.

Impact on stakeholder   1: Low Impact: no significant change. 3: Several aspects of the day-to-day operations will change. 5: High impact: Radical change.

Mind set   Describe: What guides behaviors. Interests. Motivation drivers.

Expected Resistance   Describe + 1: Strongly opposed. 3: Neutral. 5: Strongly supportive

Required Support   1: Neutral. 3: Involved. 5: Proactive - Seller.

Strategy to use   What to do: alignment, coaching, communication, training, behavior reinforcements, special resistance management

FIGURE 19.3 Template for the overall change management plan for a low-risk project (see also Chapter 15, "Understanding Your Project" and Chapter 16, "Understanding your Current Culture").

WHO SHOULD DO IT

Undoubtedly, as in any team, there will be instances of subgroup work and individual elaboration. However, the responsibility for creating the plan must rest with the executive sponsor and the team of facilitators, and ultimately with the sponsors' coalition. And, as with any plan, there will be stages of validation with other people, referents, or groups.

SUGGESTIONS FOR GENERATING THE GLOBAL PLAN

- *Aim for two goals*: to get people to accept the change and to minimize the negative impact on people—so that it does not lead to manipulation.
- *Understand that it is a single plan*, as opposed to each function or management having a piece of the plan.
- *Keep an open mind* when developing the plan. Don't fall in love with one resource or solution. Lay out the plan and be open to incorporating feedback.
- *Build frequent opportunities to review, learn, and improve* the plan itself. Install metrics and use them to check progress. Performing the activities is not the goal; it is important to achieve the intent for which they were decided.
- *First, build the strategy* and the key milestones. Then, build the plan.
- *Understand that this is continuous work.* Systematically plan to review past actions, their compliance, their impact, and new unforeseen situations.
- *Embrace improvement and positive thinking.*
- *Use feedback.* The post-training survey or the results of a Maturity for Change Survey may suggest delaying or changing actions already planned. An unforeseen event, such as a layoff, may suggest actions to counteract negative effects. A better-than-expected result might suggest a communications program to spread the word about successes and reinforce continuity.
- *Segment with a behavioral focus.* Segment the plan according to the results of the culture survey, the needs of the technical project, and the needs you encounter as the project progresses. Segment stakeholders according to the impact of the change and the reaction of the people in that group.
- *Plan for the ongoing involvement of middle management.* The feeling of disempowerment among managers is powerful.
- *Remember to include the "uninvolved."* Think not only of those who are directly involved but also of those who are not currently involved.
- *Plan to celebrate small wins.* Visibly. They may seem small, few, far from solving all the problems, but the point is to set the wheel in motion and not to kill the caterpillar—which will no longer be a butterfly.
- Quick wins have several benefits: (i) They demonstrate the true conviction, commitment, and competence of management to implement and support the project; (ii) Managers, middle managers, and employees from other areas will be able to "visualize" more concretely what is intended; and (iii) They serve as an initial experience for the facilitation team to review, learn from, and provide feedback on people's reactions to the implementation plan.

- *Plan to recognize those who suggest improvements and* opportunities, even if they are not aligned with the strategy.
- *Anchor change in the culture.* Plan opportunities to reflect on policies, processes, systems, and procedures that need to be changed.
- *Keep it simple.* "Some of the tools (methods) are really simple; they just work because we stick to them and apply them consistently."[3]

Share the plan and be open to incorporating feedback. Remember, there is no such thing as negative feedback; the negative is not taking advantage of the opportunity that exists in feedback.

Therefore, the change management plan should include activities to:

- Establish, maintain, and ensure project alignment with the organization's vision, strategy, and goals.
- Create and maintain a sponsors' coalition to support the project and the changes.
- Establish ongoing communication mechanisms that go beyond the transmission of intentions and information to ensure the reception, understanding, reflection, and consideration of employee opinions.
- Design and develop training and coaching activities segmented for each group and based on the specific needs of the project.
- Provide coaching and mentoring to key personnel to help them develop a positive outlook on how they can contribute to the change process. In particular, to help them deal with individual difficulties and losses, which are inherent to any large-scale project.
- Implement mechanisms for identifying and addressing resistance, both individual and organizational, to prevent and mitigate it.
- Implement visible mechanisms that highlight the differences between those who embrace the change and those who do not, including hiring, recognition, and promotion aligned with the intended change.
- Have a control system that allows for the monitoring and analysis of progress and results.

The following chapters will provide a more detailed overview of each of these resources.

THE MONITORING SYSTEM

It is essential to determine the indicators that will measure the impact and results of planned actions in order to (i) adjust when necessary and in a timely manner and (ii) provide the organization and sponsors' coalition with accurate information and data. Indicators may include Maturity for Change, Sponsor Evaluation, Degree of Adherence to new behaviors, Attendance at Events, Degree of Satisfaction with Recognition Actions, and Degree of Impact of Communications. In addition, technical indicators such as the effectiveness of new processes, cycle times, and quality indicators may be considered (see Chapter 27, "Metrics and KPIs," for more information).

Suggestions:

- Adhere to change by individuals and groups may change over the course of the project. The monitoring system can identify and help respond to changes.
- Include instances of evaluation of the activities of sponsors and middle management.

NOTES

1 This is based on the fact that without caterpillars there will be no butterflies. Without absurd ideas first, we will not find differentiating and attractive ideas.
2 Peter Diamandis, co-founder of Singularity University.
3 Luc Mayrand, *The Imagineering Workout*. Disney Editions, 2005.

20 Checklist for the Planning

Raúl Molteni

Note that our suggestions are based on Prosci's ADKAR model, albeit with free interpretation.

WHAT TO EXPECT FROM THE GLOBAL PLAN

A good project and change plan should include:

- *Coherence*: the culture fits the project, or the project fits the culture.
- *Honesty*: is aligned to the owners, C-suite, and management commitment, not just to the book, company, or author "of the month."
- *Entirety*: it covers design, implementation, and control.
- *Comprehensiveness*: it covers the technical, change, and project management aspects of the project.

> *Key to the outcome of the plan: culture is adapted to the project, or the project is adapted to the culture.*

RAISING AWARENESS OF THE NEED FOR CHANGE

The level of awareness regarding the necessity of the project and the desired change is crucial.

DETERMINING FACTORS

The individual's perception of the situation is shaped by their beliefs, knowledge, competencies, and previous experiences. This perception influences how they perceive the organization's problems, their own experience within and with the organization,

 DOI: 10.1201/9781003544807-26

and the credibility of the project and change promoter. In addition, the credibility of the promoter and issuer of the project and change is influenced by the rumors and opinions of referents, which may or may not align with the reasons for the project and change.

ALIGNING TO RAISE AWARENESS

Disseminate, communicate, and coach based on the vision, the why, the what for, and the opportunity for change. Include how the change aligns with the organization's vision and strategy. Integrate technical project, project management, and change management messages and concepts. Clarify the reasons for the change: what is not working—key reasons for the change—and why it is being done—goals and benefits. Explain the priority and relationship of the project and change to day-to-day and other projects. Reach out to internal audiences, not just those affected. Consider external audiences—union, market, customers, and society. Anticipate and analyze risks and resistance.

COMMUNICATING TO RAISE AWARENESS

It is the responsibility of executives to communicate effectively without delegating this task. They should engage in face-to-face interactions and consider cultural references. Walk the Talk. When developing messages, consider the vision, the "what for," the "why," the opportunity for change, the strategy, and the plan. When communicating, segment messages according to how the project impacts different groups—management and middle management require specific messages. Do not leave stakeholders unaddressed. Integrate technical project, project management, and change management messages and concepts. Consider multiple channels and continuity. Make information accessible. Remember, it's not just a matter of saying or telling. It is also about listening, reflecting, and deciding new messages.

SPONSORING TO RAISE AWARENESS

Keep the executive sponsor and sponsors' coalition active and visible to reinforce the messages. Reach out to senior and middle management. Maintain conversations to understand the impact on people and their understanding of the project. Integrate technical project, project management, and change management messages and concepts.

COACHING TO RAISE AWARENESS

Coach the executive sponsor, sponsors' coalition, and managers to better understand the change, and their individual and group roles. Correct misconceptions. Help them be prepared to deliver the communication.

HOW TO CREATE INDIVIDUAL COMMITMENT

WHAT IS INDIVIDUAL COMMITMENT?

Individual commitment refers to the initial desire someone has to participate in the project and make a difference.

DETERMINING FACTORS

The impact of the change on the individual, considering the nature and timing of the change, their personal circumstances, current situation, and future outlook. The factors that motivate the individual. The position taken by the individual's immediate superior. The position taken by other referents, including the individual's line manager, the line manager's manager, role models, peers, colleagues, and the union.

ALIGNING TO CREATE ENGAGEMENT

It is important to maintain the sponsors' coalition in a state of activity and visibility. It is essential to secure the commitment and active and visible participation of managers, middle management, and supervisors at the outset. Outline the project's priority and its relationship to day-to-day operations and other projects within the sector or group. Integrate technical project, project management, and change management messages and concepts into a unified framework.

SPONSORING TO CREATE INDIVIDUAL ENGAGEMENT

It is the owners', sponsors', and managers' personal attitude, participation, and support to the project. Plan active participation of managers, middle managers, and supervisors with their respective teams to share project objectives and the potential implications for that group. Plan to listen to staff as well as provide them with information. Communicate massively and personally at all levels, but "drill down" messages to each particular group. Emphasize the need, opportunity, benefits, relationship to strategy, and drawbacks of the project and change being addressed. Mainly, relate all that day-to-day to the people. Keep messages specific to the function or group. Be upfront about the choices and consequences for people. Listen to and address objections and prioritization of the project and change in relation to day-to-day and other projects—especially from peers, managers, and middle management. Proactively seek out and listen to objections, resistance, and comments of all kinds. Analyze causes and propose solutions to objections and resistance, especially from senior and middle management. Interact and act honestly at all times. Hold senior and middle management, and supervisors accountable.

TRAINING TO CREATE INDIVIDUAL ENGAGEMENT

Train senior and middle managers to be change leaders. Teach them to listen and learn. Deepen their understanding of the project and change. Maintain the integration of the three axes: technical, project management, and change management during the training. Train in project and change management and address the role expected of directors, managers, and middle managers.

COACHING TO CREATE INDIVIDUAL COMMITMENT

Plan to accompany them to see the benefits for each stakeholder (WIIFM—What's in it for me). Plan for personal coaching for all "bosses" and referents. Talk about and address resistance and doubts. Strengthen listening and learning skills.

MANAGING RESISTANCE TO RAISE INDIVIDUAL ENGAGEMENT

Address resistances or inhibitors that emerge from observations, conversations, work meetings, and meetings specifically convened to uncover inhibitors, among others. For example, loss of growth prospects in the organization, inconsistencies between the messages of sponsors and those of direct supervisors, contradictions between the project's change proposal and reality, and day-to-day difficulties—policies, processes, decisions, and promotions—that are inconsistent with the proposed changes. Remember that obstacles, difficulties, and resistance should be seen as positive feedback and sources of new improvements to enhance the commitments.

HOW TO GENERATE KNOWLEDGE AND COMPETENCIES

WHAT IS KNOWLEDGE AND COMPETENCIES?

It refers to the information that someone has of what they must do to fulfill their role and purpose in the project and change. This includes understanding what behaviors should change, acquiring the necessary competencies and skills, and recognizing whether the person himself/herself complies with the expected behaviors.

DETERMINING FACTORS

The level of knowledge, skills, and initial training received by the employee. Their capacity for learning. The resources available for implementing the design. The availability of time for those who are responsible for the learning and teaching processes. Access to knowledge. The budget.

TRAINING TO CREATE KNOWLEDGE AND COMPETENCIES

It is essential to clearly define the knowledge and competencies that individuals are expected to possess upon completion of the training program. To gain an understanding of the profile and initial level of knowledge of the participants, as well as their capacity to acquire the knowledge or skills that will be transmitted. Maintain the integration of the three axes (technical, project management, and change management) throughout the training process. Consider the competencies required at each stage of the project, from development through implementation of changes. The design should enable participants to:

- Gain a full understanding of the intended outcome.
- Connect the training content to their day-to-day work experience.
- Translate the new knowledge to their specific work context immediately following the formal training.
- Determine whether the results achieved align with the expected outcomes.

The design should incorporate the superior's guidance both before and after the training period. On-the-job training should lead to quick wins.

Coaching to Create Knowledge and Competencies

Accompany to solve the doubts that arise in the daily application. Plan support to see "a single project." Accompany to see the relationship between personal development and alignment with the expected benefits of the project. Support to achieve quick wins.

Resistance Management to Create Knowledge and Competences

Address resistance or inhibitors that arise from observations, conversations, work meetings, meetings specifically convened to identify inhibitors, difficulties in applying what has been learned, and difficulties in achieving quick wins, among others. Remember that barriers, difficulties, and resistance should be seen as positive feedback and sources of new improvements.

HOW TO GENERATE CONTINUITY AND ROUTINES

What Is Continuity and Routines?

The ability and capacity of someone to consistently and sustainably apply the expected behaviors and habits of the project and change. Execution with quantifiable results.

Determining Factors

Incomplete learning. Lack of support from supervisor. Lack of necessary resources. Overconfidence generated by a quick—and superficial—understanding of the new habit or behavior. Psychological blocks. Physical or cognitive limitations in reaching the required level of competence. Limited time for development. The training is considered as a "one shot" type. The level of importance and specificity of the recognition given to those who show commitment. The relationship between reinforcement and the level of compliance.

Sponsoring to Create Continuity and Routines

Maintain the visibility of the sponsors and the sponsors' coalition, especially with managers, middle managers, and direct supervisors. Solve inhibitors. Work with them to address barriers and resistance. Install integrity and compliance indicators to standardize new behaviors. Use indicators. Ensure resources and control of project indicators. Encourage people to suggest and identify what needs to be changed. Align incentives. Promote the project to new people, areas, and units.

Coaching to Create Continuity and Routines

Listen to all. Promote individual and group discussions to analyze the evolution of the project. Identify the sources of resistance and obstacles. Implement actions to remove the obstacles. Use the indicators. Plan individual and group support to:

- Eliminate any doubts about the project and its evolution.
- Resolve the doubts that arise and maintain continuity in the application of new behaviors.
- See "One Single Project".
- See the relationship between personal development and alignment with project benefits.
- Resolve outcome inhibitors.

Training to Create Continuity and Routines

Use the indicators. Review, redesign, repeat, or complete the training required to implement knowledge and skills.

Resistance Management to Create Continuity and Routines

Use the indicators. Address resistance or inhibitors that arise from observation, conversations, work meetings, or meetings specially convened to reveal inhibitors. Identify difficulties in applying what has been learned and achieving results. Identify sources of conflicts and analyze the results of the work environment, among other things. For example, there may be a lack of sustainability or consistency in resource allocation, a contradictory attitude on the part of the direct manager, nonexistent or inadequate recognition, or a lack of measurement of results.

HOW TO GENERATE LEADERSHIP AND ORGANIZATIONAL COMMITMENT

What Is Leadership and Organizational Commitment?

Consistency between the vision and the need communicated at the beginning of the project, and the additional and systematic efforts of the change promoters to remove inhibitors and continuously improve the project, sustain the change and make it part of the culture—anchor it.

Determining Factors

Loss of coherence, agreement, and teamwork among the sponsors' coalition. The results did not meet the initial expectations. The project was the result of an "inspiration of a summer night." A new project is "discovered" as the new "savior." The perception—I emphasize "perception"—that the stakeholders have of the commitment of the management and/or the promoters of the change. The presence or absence of a continuous project improvement process. Not measuring or ignoring indicators.

Aligning to Create Leadership and Organizational Commitment

Give visibility to the change promoters—owners, directors, CEO, or unit head, depending on the type of project—to support the project as if it were just starting. Have them own the results of the project. Maintain and make visible the relationship between project results and benefits to the organization and stakeholders.

Sponsoring to Create Leadership and Organizational Commitment

Provide visibility to change owners, directors, CEOs, or unit managers to participate in project analysis discussions, celebration, and recognition. Establish public recognition systems that are fully aligned with the intended change. Encourage one-on-one and group conversations and approaches at the executive and middle management levels. Use indicators. Plan to have them participate actively and visibly in analyzing results and generating improvements. Enable second generation governance.

Coaching to Create Leadership and Organizational Commitment

Accompany with visibility to owners, directors, CEO, and sponsors' coalition to analyze results, inhibitors, and resistance and learn.

GENERAL SUGGESTIONS FOR DEALING WITH RESISTANCE

- *Be honest.* Don't hide what everyone will eventually see—and see badly. How much of the change—or part of it—is nonnegotiable? What is defined and what is not? Don't minimize or try to hide losses or difficulties.
- *Communicate.* Honest and consistent information about what is intended. Engage in face-to-face conversations to listen to and understand others' opinions and positions about the project, the changes it will entail, and the reasons and emotions it evokes in others.
- *Provide opportunities to participate.* Involve. Systematically. About the project, about the change, and about one's own possibilities and actions. Without forgetting to "involve those not directly involved."

- *Create opportunities for action.* The project must provide concrete instances to apply new knowledge and behaviors and thus be able to see the mutual benefits—organizational and personal.
- *Be patient.* Change doesn't happen immediately; it takes time to assimilate, understand, partner, and participate. Conversations provide a sense of time and positive feedback. Patience does not mean giving in or accepting the status quo.
- *Avoid surprises.* While surprise gifts are welcome on birthdays, this is not the case in change processes. Surprises should be avoided during projects.
- *Limit the change to what is necessary.* More aspects undergoing change can delay the generation of awareness and commitment, as well as open new fronts of discord. Less is more.
- *Keep the change leader visible.* Depending on the nature and scope of the project, the most senior leader of the unit should be visible and accountable.
- *Exercise empathy.* Put yourself in the other person's shoes before judging. Look for win-win situations rather than pigeonholing people as pro- or anti-change. Allow for different points of view—remember the 2-6-2, allow for the two opposing factions and look out for the six who are in doubt or expectation.

Understand your interlocutor!

MAIN ASPECTS

Promote, require, monitor, and disclose improvements to policies, processes, and procedures that are inconsistent with what is being promoted and required by the project and change. For example, changes in promotion and compensation policies.

Responsibility of the person driving the change—owners, directors, CEO, or unit manager, depending on the nature of the project—is : Promote, require, monitor, and disseminate the improvement of policies, processes, and procedures that are inconsistent with what is being promoted and required by the project and change.

21 Clarity and Alignment

Raúl Molteni

WHY A PLAN FOR CLARITY AND ALIGNMENT?

Design and execute actions that keep the project aligned and contribute to the goals and strategy that gave rise to it and that are consistent with those of the organization. See Chapter 20, "Checklist for the Planning," for suggestions.

THE PROJECT'S WHY AND WHAT FOR

Whether they realize it or not, the leaders who truly inspire people do so by following a naturally occurring pattern that Simon Sinek calls "The Golden Circle."[1] The Golden Circle provides compelling evidence of how much more we can accomplish if we begin everything we do by asking a simple question: "why?" It finds order and predictability in human behavior. Simply put, the Golden Circle helps us understand why we do what we do. Your why is your purpose, cause, or belief.

It is also needed to be aware of the "what for." To mobilize the whole organization—or a great part of it—we need something more: projecting us into the future, having a vision, a better "us." There are two straightforward statements with a significant objective to achieve them:

- *Understanding the what for and the why of the project*—in relation to the strategy. The what for of the project will guide and condition the tactics, the approach, the tools, and the practices that support it. The why[2] will remind us of our character to do it.
- *Align the organization*—strategy, policies, objectives—with a consistent project.

ALIGNING EVERYONE

It is reasonable to assume that we will not be able to commit an entire workforce to the desired change, given the numerous variables that condition it, many of which are

DOI: 10.1201/9781003544807-27

beyond the organization's control. The objective is to identify and engage a critical mass of individuals who can drive the desired change.

The kickoff is not the starting point. **Before initiating change, it is essential to achieve critical executive mass**. Once initiated, aligning the entire organization requires a continuous and ongoing effort, like the commitment required for effective communication. This is an activity that requires dedicated planning, execution, and control time from executives and the sponsors' coalition:

> Involve everyone: stimulating the universal involvement of all individuals in an organization, creating ownership and capabilities for assuring the quality of their own work and in making improvements endlessly.[3]

ALIGNING EVERYTHING

In addition to considering the people and functions involved, it is essential to assess the alignment of policies, systems—not only IT—and processes. It is essential to consider not only those directly related to the change but also those indirectly impacting it.

I conducted an in-depth investigation with the improvement teams of an assembly plant to identify the root causes of the "missing parts" issue, which resulted in products remaining unfinished due to a lack of necessary components. As a result, the process that was intended to conclude within the plant was concluding in a parking lot outside the plant, and several weeks later. The five whys and process mapping techniques enabled us to identify the root cause of the problem, which led us to the suppliers, from there to the purchasing process and, finally, to the cash flow usage policy. The underlying issue was not directly related to the process itself, yet its impact was still felt.

By "clarity and alignment," I mean a coherent and profound action that aligns with the vision and aspirations of the project. Some questions to check for consistency are as follows:

- Are the plans and projects of all functions aligned with the vision?
- Is the organization's capacity—resources and staffing—consistent with the vision?
- Are information systems ready, or in the process of being ready, to sustain change?
- Is the economic-financial policy aligned with and supportive of change?
- Is the HR policy aligned?
- If they are not, is their alignment somehow foreseen and decided?

Most projects start without much clarity about what change they are trying to achieve. Understanding the why and what for the project is key.

WHO AND HOW

The primary responsibility lies with the sponsor and/or sponsors' coalition. They are the ones who must ensure that the project is aligned with the organization's strategy and goals and stays that way. They are also the ones who must ensure that the various components and subprojects are aligned with each other and with the project itself.

The Facilitation Team provides tactical resources for the installation, validation, and alignment of processes. To prevent deviations, it is essential to maintain focus on the intention, purpose, and rationale behind each action in the plan. It is not sufficient to assume that an action has been completed successfully just because it has been executed. The purpose of communication is to provide information, facilitate understanding, and collect feedback, not to label a Gantt action.

Implementing pulse surveys and holding regular meetings with staff at all levels and functions involved in the project are key tools for assessing whether tasks have been completed and objectives achieved.

> *The success of the project will depend on the quality of the actions and interactions, not just the quantity of actions executed.*

NOTES

1 https://simonsinek.com/golden-circle/
2 The why as understood under Simon Sinek's Golden Circle meaning.
3 Narayanan Ramanathan and Gregory H. Watson, *Quality Manifesto for the 21st Century*. International Academy for Quality, 2021. www.ihi.org/resources/Pages/Improvement Stories/ThePowerofHavingtheBoardonBoard.aspx. https://iaq.wildapricot.org/resources/ Documents/Position%20Paper/Revitalizing%20Quality%20-%20The%20Quality%20Ma nifesto%20for%20the%2021st%20Century%20-%202021.pdf

22 Communication

Raúl Molteni

Key to all messages: Keep it simple and honest. It's better to find out something doesn't work as planned than to find out it was a lie.

WHY A COMMUNICATION PLAN?

To create spaces of understanding between the perspectives of the project—those of its promoters and leaders—and those of the staff—both at the individual and group levels. This is what is expected:

- Information and mutual understanding to gain Awareness, Commitment, Knowledge, and Continuity with the project.
- Space to understand each other's positions, emotions, reactions, and motives.

CONTINUITY OF COMMUNICATION[1]

It is not just an initial plan. It is a plan with ongoing updates and feedback to help people move from Phase to Phase, from Awareness to Commitment to Knowledge to Continuity.

STEPS TO ADDRESS COMMUNICATION

I recommend the following steps for outlining the various communication channels.

1. *Identify the messages to be delivered and their purpose.* Is it to create awareness of the need for change? Commitment? Knowledge of what needs to be done? Continuity of change? The answers will help determine who, when, and what to communicate.
2. *Identify audiences, key messages, and media.* There will be times and messages that are appropriate for all stakeholders, while there will be others that should—not just can—be targeted to segments of specific groups:

DOI: 10.1201/9781003544807-28

- The "uninvolved" parties. Groups of people who, due to their involvement in other areas, functions, or units, do not appear to require any information or communication. Keeping them informed could help to prevent the subsequent spread of rumors that are detrimental to the project and its changes.
- Customers and/or suppliers. The implementation of changes in agile cells or design and improvement projects will have an impact on them.

3. *Generate content*: frequencies, formats, methods, dates, people responsible. Think about the audience's WIIFM,[2] not just what you want to say.

4. *Obtain feedback* from people not involved in the design process and secure approval from relevant stakeholders—including both internal and external communication teams.

5. *Define how information needs that arise daily will be managed.* Prepare the process and methodology for sudden communication needs (ad hoc communication).

> *Prepare the process and methodology for dealing with suddenly emerging messages (ad hoc communication).*

HOW TO IDENTIFY WHAT TO COMMUNICATE

There is nothing wrong with thinking about what you want to communicate. But at some point, you need to pivot and listen to your stakeholders to understand what you need to communicate, not just what you want to communicate.

When validating the results of a communication activity, it is common to find expressions such as "I didn't understand well," "I don't remember anything in particular," "There was nothing new," or "Something is not good about the change." We could have given a thousand messages about the project or about the different projects that were going on, which led to a certain saturation of the employees. Or people can't see the logic of what is being communicated. It's not their fault or yours. It is just a new need for communication.

When several projects coexist in the organization—a more than usual situation—communication must be designed integrally. The fact that the projects have different objectives and are promoted and sponsored by different managers does not justify the lack of coordination and integration.

> *The fact that the projects have different objectives and are promoted and sponsored by different managers does not justify the lack of coordination and integration of communication.*

GUIDANCE FOR IDENTIFYING MESSAGES THROUGHOUT THE PROJECT

Suggestions for Raising Awareness

The communication plan should address the kickoff, if any, preferably with directors, CEOs, sponsors, and managers as senders. The following suggestions also apply to subsequent stages as the change unfolds:

- (i) What is the project for? (ii) What is the initial situation? (iii) What is the relationship with the organization's strategy and objectives?
- (i) What is intended? (ii) What are we doing this project for? (iii) Why now? (iv) What would happen if the changes were not made? (v) What are the expected benefits?
- (i) What is the plan of action? (ii) What are the stages? (iii) What is the scope?—policies, functions, structure, technology, processes, and relationships.
- (i) How will it affect the different stakeholders? (ii) What changes will each person see? (iii) What things should not change?
- (i) How long would it take to see results or differences? (ii) What results should be seen?
- (i) What are the risks of the plan?

Messages for Achieving Commitment

Preferably using senior and middle managers as the senders, the following are suggested messages to gain commitment from those who will or should participate:

- (i) What will change for each person? (ii) What is the impact on his/her daily work?
- (i) What will it require him/her to change? (ii) What will it require him/her to maintain on an ongoing and systematic basis? Specifically, what behaviors will he/she have to change, abandon, or adopt?
- (i) What benefits will it bring to him/her? (ii) What's in it for him/her?—from the employee perspective.
- (i) How have other organizations made similar changes?

Messages to Gain Knowledge

These are suggestions of basic messages for communication in cases of the acquisition or the improvement of knowledge and skills, with teachers and specialists with validated authority as senders:

- (i) What is the purpose of this training activity? (ii) What are the specific behaviors, skills, or routines to be consolidated?
- (i) How will it be developed? (ii) What does it specifically require of the person—commitment, attention, effort?

- (i) What is expected at the end of a particular activity? (ii) What happens after the activity? (iii) What do they want to see? Who and when?

MESSAGES FOR ACHIEVING CONTINUITY

These are basic message suggestions for communication to achieve or improve continuity. All senders count for this purpose.

- (i) What activities have already been completed? (ii) What was their objective? (iii) What has been achieved? (iv) Why is this important in relation to the purpose of the project?
- (i) What are the main benefits achieved so far?
- (i) What are the main obstacles encountered at present?
- What do people think about the project? What obstacles do they see? What are the inconsistencies? What do they see as positive?
- Share what is being done or planned to. For example, to identify the cause of resistance common to many employees, the lack of certain resources, to review policies and processes, to remove an obstacle found in a function.

SUGGESTIONS FOR MESSAGES TO DELIVER

Here are some more suggestions from the field:

- *Briefness.* Less is more. Clear and short sentences. If we can find a catchy slogan, all the better.
- *Honesty.* If our words are not credible, or if we are not credible, our interlocutors will not believe the message and will not buy it. Moreover, they will certainly damage the image of the project, the company, and themselves.
- *Simplicity.* Speak in the same language as your audience. Avoid jargon and complicated terms that require the audience to pay close attention to understand.
- *Inspiration.* Not every message developed from logic will be accepted or understood. Consider emotions. It is not just a matter of presenting the logic of change but of influencing people's emotions, which will then guide their actions.
- *Visualization.* A mental image can help more than many words. An identifiable visual slogan is more likely to fix the message in the audience's mind. I mean an image of the message, not meaningless pictures or drawings.
- *Understanding.* Communication is not information. Let's inquire, listen, ask questions, and understand what people understand and how they perceive the messages.
- *Contextualizing messages.* Always relate the why, what for, and how of the message of the moment to the main change and to the business. And most importantly, to what they are experiencing.

SUGGESTIONS FOR LISTENING

Listening is an art, as Bradlee Snow[3] said. A fundamental art of communication. For understanding others and of ourselves. I mentioned in Chapter 6, "About Continuous Communication," that communication is not about transmitting messages. It is about understanding others. Therefore, listening is just as important as sending. Start by listening. There is nothing wrong with thinking about what you want to communicate; however, at some point, you need to pivot and listen to your stakeholders to understand what you need—not just want—to communicate.

Ask

Asking naively carries with it the intention to listen and understand what the other person is trying to say. Some suggestions:

- State the purpose of the conversation.
- Explain the reason for asking questions.
- Use synergistic language, to understand the emotions and reactions it may provoke in the other person.
- Use a combination of open and closed questions.
- Ask for the reasons behind the answers, especially with data.
- Consider the other person's answers to structure the question sequence. An answer should suggest the next question.

Paraphrase

Express to the other person what you have understood in your own words. Check with him/her to see if your understanding matches the other person's intent. Often, letting the other person "hear themselves" leads to clarifications that improve the conversation and understanding.

Pleadings

State your thoughts and your opinion about the situation. It can help:

- Explain using premises, reasoning, and data.
- Use examples, not anecdotes.
- Point out when opinions are unfounded or based on emotion.
- Encourage the other person to ask questions to clarify what has been said based on doubt or understanding.

THREE INEFFECTIVE PRACTICES

These are practices I see often though they never achieve successful results:

- Very good vision and message design, but poor communication.

- A few people, including the CEO, are visibly committed to communication, versus a majority of managers and bosses who, like in a bullfight, shy away from involvement.
- Very good design of vision and messages, systematic activities, and visibility of middle management. With day-to-day actions that go against what the project proposes but are nevertheless maintained. Communication is not only verbal; behaviors, policies, processes, and procedures also communicate.

IMPROVE THE COMMUNICATION PLAN

Within the plan, activities that provide feedback about the communication plan itself should be considered. For example, activities to measure how stakeholders perceive and "live" the communication and what they understand and learn from it.

Do not forget to provide feedback about mechanisms to understand whether the communication is—and has been—effective.

NOTES

1 See Chapter 20, "Checklist for the Planning."
2 "What's in it for me."
3 Disney Imagineers, *The Imagineering Workout*. Disney Editions, 2005.

23 Education and Training

Raúl Molteni

WHY AN EDUCATION AND TRAINING PLAN

A training system that minimizes the resistance that results from people not knowing, not being able to do, not being able to say, and not being able to hear what they are being asked to know, do, say, and hear. And that eliminates, as much as possible, the stress, distress, and difficulties that people experience because of not knowing or being able to do what they are being asked to do as a result of the project.

ABOUT THE PLAN

To be effective, training requires that the recipient be committed to and on board with the project. This person will seek out the new knowledge that will enable him or her to manage the change successfully.

The training plan should include activities with stakeholders, both individually and in groups, to provide them with the knowledge, skills, and competencies they need to perform the new behaviors targeted by the project.

> *Training is effective when the trainee is committed and on board with the change.*

They may be technical—such as how to use a computer system or a measuring instrument—or social—such as how to work more effectively in a team. Or they may be skills designed to mitigate a particular resistance.

We must not lose sight of the importance of support materials, the selection and preparation of coaches to provide support, and a back office for consultations.

STEPS IN CREATING THE TRAINING PLAN

It is a plan with constant updates and feedback as people move from Awareness to Commitment to Knowledge and Continuity. It is the HR area that needs to work

DOI: 10.1201/9781003544807-29

intensively on this aspect, although the training of Green and Black Belts, Scrum Masters, or other roles may be predetermined in some way.

See Chapter 20, "Checklist for the Planning," for suggestions.

ABOUT THE PROJECT'S NEEDS

It is to analyze the needs of the project—technical, social, and project management—in terms of knowledge and behaviors. For example, it is to ask what is expected of a Black Belt in this organization after the change. It is not mandatory to train according to the content of a known body of knowledge.

What is expected of this role in this organization from the project?

ABOUT THE KNOWLEDGE AND SKILLS REQUIRED

It is essential to clearly identify the knowledge and skills that the new behaviors require in all aspects. For instance, what kind of competencies regarding the project does each sponsor need to have? What knowledge and skills should the Black Belt possess? What knowledge of Lean should he or she have? Will he or she require any support, for instance, from HR, if he or she finds it inconvenient to facilitate a team of Green Belts? It is also important to consider the knowledge and skills that will eventually be needed for the transition period between the current situation and the one desired by the project.

ABOUT THE CURRENT KNOWLEDGE AND SKILLS

It is essential to identify the specific knowledge and skill profile of the individuals to be trained. For example, what kind of competencies regarding the project does each sponsor have? Will the Black Belt require training from the beginning, or does he/she already have Green Belt training? Does the candidate have project management knowledge and experience? Does the candidate have experience facilitating teams? Does the candidate have experience in change management? What kind of competencies regarding the project does each sponsor have?

ABOUT THE GAPS

Determine the training requirements based on the discrepancies between the current profiles and those required by the change. Both those resulting from the intended change—future situation—and those required for the transition period. For example, what knowledge or skills will help the Black Belt get through this period with less uncertainty, effort, and stress? And when will he/she lead a project?

On Designing the Agenda

Design training to integrate responses to all types of gaps. Design for need and avoid generic programs that may include a lot of theory but may not help meet the required behaviors.

> *Avoid generic programs that may include a lot of theory but may not help achieve the required behaviors.*

Following trends can be positive, but not always. I suggest that you challenge yourself when choosing training dynamics and not get carried away with techniques that seem modern but do not contribute to the goal. Participatory games that generate integration are motivating, but they can be detrimental to the focus of the training.

Selecting and Designing Resources

This involves determining the agendas, coordinators, materials, and resources needed for the training. For example, will the activities be virtual or face-to-face? If so, what resources will be used? What is the profile of the coordinator? External or internal? Who will monitor the Black Belt's level of understanding, and how? How much time will be required?

ON TRAINING COORDINATORS

It is essential to prioritize the output and ensure that a single coordinator is not expected to handle all topics. It is also important to ensure that the training is not left in the hands of a group of instructors who have not been coordinated. Teams of coordinators should be prepared with a "train-the-trainers" approach, ensuring not only technical but also academic competencies.

For example, is there anyone in the organization who is competent to train Black Belts? Who will own the change management and team facilitation agenda? Will the profile of the coordinators you are considering be accepted as authoritative by managers?

On the Job Training

Since we do not all learn in the same way, it is to be expected that, no matter how well designed and coordinated the training, not all participants will acquire the knowledge and skills to the extent required. On-the-job training is an important part of training and should be planned and structured in the same way as "classroom" training. Managers, supervisors, and line managers need to be prepared to provide on-the-job training for their employees.

Managers, supervisors, and line managers need to be prepared to provide on-the-job training for their employees.

ON TRAINING DEVELOPMENT

Training should be developed in an integrated manner, avoiding differentiated courses by topic that disintegrate the vision and behavior of the participants. Monitoring in real time the participants' understanding of the topics helps to understand whether they see the knowledge in relation to the project and their future day-to-day work, or in relation to several projects proposed by different areas and with different objectives. For example, will there be a concurrent project development to facilitate learning by doing?

Will there be a simultaneous project development and training to facilitate "learning by doing"?

ABOUT VALIDATION

Training evaluation should be part of the design. The training activity is not the goal; the acquisition of knowledge or competence by the participant is. Taking an activity for granted, with the signature of the participants and the trainer, is a common "trap" for an auditor, but useless for the project. Exams and certifications are part of this stage, but the independence and authority of those who perform and approve them must be assured.

For example, how would you validate the knowledge acquired? How would you validate that the Black Belt has acquired the necessary skills to apply techniques such as correlation techniques? How would you validate that he/she has the minimum skills to facilitate the project team? How will you validate that each sponsor of the coalition has the basic skills to support the project?

Taking an activity for granted, with the signature of the participants and the trainer, is a common "trap" for an auditor, but useless for the project.

ABOUT THE DESIGN AND DELIVERY OF SUPPORT

Preparing agendas, materials, and support for the next level of training should be part of the overall design—see Chapter 24, "Continuous Support." For example, is there a trained Master Black Belt available to monitor and support the newly trained Black Belt? Who will help the sponsors' coalition, with reflection and analysis, to effectively fulfill their role?

Who will help the sponsors' coalition, with reflection and analysis, to effectively fulfill their role?

SPONSORS' COALITION, SENIOR AND MIDDLE MANAGEMENT, AND FACILITATOR TEAM

They are the ones who should be trained first. I mean serious training, not an introduction that does not validate new competencies. See Chapter 13, "The Sponsors' Coalition," and Chapter 8, "About Owners, the Board, and CEOs."

TRAINING AS EVIDENCE OF MANAGEMENT COMMITMENT

An additional consideration is the allocation of budget and staff time for training. This is clear evidence of management's commitment to the project.

A medium-sized company has embarked on a transformation process to face the future. Its CEO and major shareholder define this process as key to the organization. As part of this process, the CEO himself deems it necessary and launches a development project for all his directors and middle managers. A cultural survey and an individual assessment confirm and clearly show the training deficit of those who occupy these levels. He wants to see behavioral results in less than six months but decides that middle management should only spend two hours per week on their training. Is this time congruent with the objective?

If you think training is expensive and time-consuming for those who attend, try ignorance (based on W. E. Deming's phrase).

SUGGESTIONS FOR LEARNING

The learner plays a crucial role in the learning process. It is challenging to provide instruction to an individual who is not engaged in the learning process. As a learner, there are a few key suggestions to keep in mind on a permanent basis:

- *Be curious.* To inquire. To observe rather than to view. Do not make assumptions.
- *Ask "what for."* Take note of the response. Take a moment to reflect. Develop the concept further.
- *Understand that listening is not hearing.* Reflect on what was said. This is to evolve the conversation.
- *Maintain self-criticism.* Understand that we know much less than we think we do. And we know even less about what we don't know.
- *Empathize.* To adopt the perspective of the other party. To act as the other party. To respond emotionally to the other party.
- *Leverage* your own experience.

ABOUT COMMON PRACTICES

- *Ensure design facilitates learning through hands-on experience.* Knowing the theory that supports a practice is necessary, but the aim of the project goes beyond that; it is to improve and transform.
- *Prioritize learning over fun.* Ensure that the dynamics are engaging and enjoyable but also conducive to learning.
- *Focus on prior awareness and commitment* by those who are key influencers for the personnel.
- *Consider personalized design.* With content and pace tailored to the learner's capacity.

24 Continuous Support

Raúl Molteni

THE PURPOSE OF CONTINUOUS SUPPORT, WHAT, AND WHAT FOR

Provide individual and group support to gain commitment to the project, to better and more easily apply the knowledge and behaviors required by the project, and to identify and eliminate inhibitors that may arise during the Continuity phase. Operationally:

- To establish a system of relationships and conversations between referents for the project, and with professionals prepared to assist them to identify their role within it.
- Facilitate discussions with staff to identify the emotions and conversations that influence their perception and interpretation of the project based on their beliefs and experiences.
- Analyze the positions and reactions, both positive and negative, that these emotions and conversations elicit.
- Develop strategies to address the concerns and reservations that staff may have about the project and find mutually beneficial solutions.

CHARACTERISTICS OF ORGANIZATIONAL COACHING AND MENTORING

In terms of organizational coaching, which forms part of the activities to be implemented for transformation, I envisage the following characteristics:

- *Two objectives*. One for the organization and the other for the people.
- *It may be optional*. Nevertheless, it is essential to gain a deeper comprehension of the project—and the role within it—that has proven to be strategic for the organization.
- *The conversations are not freely chosen by the coachee*. They focus on the relationship between the coachee and the project. The goal is to accompany the coachee so that he/she can see the project for himself/herself and without

DOI: 10.1201/9781003544807-30

bias, reflect on his/her place in the project, and share doubts and risks related to his/her participation. It is support focused on the specific needs of each person, on their personal scenario, on their role, on their doubts and fears, and on their dilemmas related to the project.

- *It is to accompany and help the coachee to see himself/herself in the project* and to find ways to overcome it with the greatest possible benefit—for him/her and for the organization.
- *It is a conversation with a clear purpose*, about the person's relationship with the project.
- *It is limited to the time needed* to achieve the goal. It is not an endless activity.
- *It is more than containment*, although it may serve as such. It is a conversation to resolve doubts and find new ways of thinking, understanding, and solving.
- *The main benefit I see in these activities is learning to learn.* Conversations that allow reflection on how to use individual and organizational experience for the project and for future benefit. As Peter Drucker put it, "We now accept the fact that learning is a lifelong process of keeping up with change. And the most urgent task is to teach people how to learn."

ABOUT COACHES AND MENTORS

It is crucial for the coach to establish a sense of empathy with the coachee and foster a climate of complete trust. The selection of the coach for a specific coachee is really important. It is not simply about "the coach we have" but rather about identifying "the coach we need."

STEPS FOR CREATING THE COACHING PLAN

The coaching plan should address the actions that need to be taken with the various stakeholders, both individually and in subgroups, to accompany and support them in reflecting and acting on their participation in the project and its impact.

THE STEPS

1. Identify the people and groups who need and would benefit from this type of conversation based on the timely impact assessment—see Chapter 18, "What Could You Be Facing?"
2. Formalize and prepare the coaches. The conversations are not "coffee conversations"; they are methodical conversations. The coach's questions are designed to help the coachee see how his/her experiences, beliefs, and values shape his/her opinions about the project and his/her participation. It will help him/her not to be tempted to suggest solutions, positions, or decisions before provoking the coachee's reflection; the coachee's relationship with the project and his future in it should be "discovered" by the coachee. As part of the preparation, the coach should ensure that the coachee has the best possible understanding of the "what for," "why," and "how" of the project.

3. Develop agendas for both individual and group sessions, identifying key topics that align with the profiles and behavioral objectives.
4. Prepare coachees. One common question is, "why me?" The purpose must be clearly communicated: these conversations are designed to facilitate reflection on the relationship between the project and that individual. It is important to note that this is not an evaluation nor a penalty.
5. Distribute the agenda to each coachee.
6. Implement a system for monitoring progress. As with any indicator, our focus is on the intention, not just the realization of the plan. Therefore, what needs to be monitored is (i) whether the coachee is maintaining a better understanding and relationship with the project, not just whether the conversations are taking place, and (ii) whether the coachee is adding competencies for his/her better and greater contribution.

See Chapter 20, "Checklist for the Planning" for suggestions.

IN THE CASE OF MENTORING

While the suggestion to the coach not to give an opinion is clearly a position, it is also important to consider the option of doing so. When there is a pressing need to attain the competencies or level of participation of the referents, lack of expertise and preparation on the part of the coachee, or high complexity of the objective, the role of the mentor becomes crucial. The mentor leverages their experience, expertise, and knowledge to guide the mentee.

In our case, coaching and mentoring, although different, maintain the same principle: guiding learning with questions that show situations from a different perspective than the person can see without collaboration.

25 Managing Resistances

Raúl Molteni

WHY A RESISTANCE MANAGEMENT PLAN?

To create a system that facilitates the identification, recording, and subsequent action to anticipate, eliminate, and mitigate the causes of resistance or to minimize its effects. Those that have not been considered, those that have arisen because of the project's development, or those that have not been properly mitigated.

See Chapter 20, "Checklist for the Planning" for suggestions.

WILL THERE ALWAYS BE RESISTANCE?

Expect resistance. Reactions to change are emotional and instinctive. It is the limbic system responding to a threat.

The resistance people express to our proposal—as well as the resistance we express to the proposal of others—has to do with a natural reaction of our brain. It depends on our own factors: our emotions, our experiences, our knowledge, our sense of controllability, and our perception of our possibilities. And on external factors inherent to the project itself and to the conditions under which it is proposed. Resistance is not only verbal. Absenteeism, high turnover, low efficiency, and lack of continuity in the application of new procedures and routines are also "speaking" resistance.

Harwood Manufacturing Corporation's experience many years ago showed something that still holds true today. The study found that 38% of operators who had to adapt to new procedures returned to their pre-change performance levels. The other 62% maintained levels below the new standard or changed industries. It also showed that the time required to retrain an operator was longer than the time required to train a new operator. This experience led to two conclusions:

- Resistance to change cannot be avoided, nor can its consequences. But it can be dramatically reduced.
- People's participation in the changes is a significant and positive variable.

DOI: 10.1201/9781003544807-31

- How and when to assess compliance with a change should be carefully considered.

THE PURPOSE OF RESISTANCE MANAGEMENT

We manage resistance to:

- *Minimize the impact of noncompliance and delays* associated with employees' resistance to applying the new knowledge, skills, and behaviors required by the project.
- *Minimize the discomfort, conflict, attrition, and stress* generated by personnel because of their misalignment with the project.

We must prepare ourselves to anticipate, categorize, and deal appropriately with resistance. Unanticipated resistance will always exist and, therefore, requires special and unavoidable preparation. A good strategy and plan will address the prevention and mitigation of those resistances that are anticipated, as well as plans to identify and address unanticipated resistances as they arise.

TYPE OF RESISTANCE

We can have physical and cognitive activities. Resistance comes not only from changing physical tasks but also from cognitive ones. Cognitive processes are those by which we take in information, pick it up, play with it, analyze it, assemble it, reorganize it, judge and reason with it, make inferences, plans, and decisions, and take action.[1]

These kinds of resistance are common:

- *Ideological.* This postulates concepts in favor of acquired rights and working conditions—generally coming from the sphere of labor or union representation.
- *Conceptual.* This will be related to different business approaches.
- *Operational.* The question "What does this mean for me?" is postulated.

While all should be considered, priority should be given to those that have the greatest impact on the project.

THE KEY—UNDERSTANDING THE CAUSES OF RESISTANCE

The key to preventing or mitigating resistance is to understand its cause. The exposed argument doesn't necessarily reflect the true cause. In the same way that really understanding the needs and expectations of customers requires more than asking them, the comprehension of the real cause for resistance requires more than asking a person. Real communication, paying attention to what happens and will happen after the change with the person in his day-to-day operation, and understanding his/her aspirations are a must if you want to tackle the cause.

Root cause analysis—using five whys, for example—or similar techniques is a very good means. Facilitators with technical knowledge can help those with social knowledge to comprehend and design effective countermeasures.

THE MOST COMMON CAUSES OF RESISTANCE

It is worthwhile to summarize the most common causes of resistance:

- *Lack of information about the change*. There is little to understand about the project itself and the changes it implies for employees.
- *Increased workload* and limited time availability. Given the specific characteristics of the project, there is a lot to do, and it does not seem to be feasible given the amount and availability of resources.
- *Change saturation* given the number and variety of simultaneous changes in the organization. Management may view these as independent, and therefore, the saturation is masked. Functions such as IT—everyone is asking for changes in systems—HR—everyone is asking for training of some kind—or Production—everyone's asking for something from supervisors—are usually the most obvious, but not the only ones.
- *Personal characteristics or situations*. Such as fears related to the loss of "power," uncertainty about the future—job stability, development, permanence in a function or hierarchical level, the potential loss of compensation or loss of status quo—and bad experiences with past changes within the current or other known organizations.
- *Fears*. For speaking up, making mistakes, being judged, or saying no.
- *Interests that are not aligned with those of the organization*. Such as blocking a repositioning project of the company because of a rupture and conflict between two parts of the union.

HOW TO IDENTIFY RESISTANCE

Resistance to a project, to a change, to a new idea is natural. It is not a reflection, at least not 100%, of the quality of the project or plan. It reflects what these people say to themselves, their emotions, and the state of mind created by their internal conversations—coming from their beliefs, experiences, knowledge, and what they hear, especially from their referents. The key is to examine and understand how they relate to the project. Some questions to help you understand resistance are:

- What and how much do you know about the other person?
- Are you sure he/she understood you?
- Are you sure you understood him/her?
- What role and motivation does he/she have?

- Do you understand what memories it brings back to him/her and what emotions the project evokes in him/her?
- What is his/her argument, reasoning, and suggestions? What part of those are reasonable and what can you learn from it?

How

- *Personal relationship*. This is my preferred medium. Face to face. The product of a confidential conversation. It could be an informal conversation about an operational problem, an Agile cell work meeting, a data correlation analysis in a DMAIC,[2] or a meeting called to talk specifically about how the person feels about the change. It could be a preplanned coaching conversation or a comment in a training activity. The key is to have a willingness to listen and see cues—gestures, tones of voice—to register and channel. Basically, it's being on Gemba.[3]
- *Focus groups*. It would be worthwhile for these activities to be part of the change management plan to check the opinions and emotions that the project is generating in different stakeholders.
- *Change indicators*. These indicators—see Chapter 27, "Metrics and KPIs"— show the evolution and, therefore, make it possible to identify aspects of the change that require action.

Who

Anyone who is involved in the change in any way can bring resistance and discomfort of any kind to the surface. A coordinator coaching PDCA,[4] a Scrum Master[5] after a Sprint Review, or the supervisor of an operational unit. And, of course, the sponsor and the facilitator team members. The important thing is not so much the who but to have a means available so that the opportunity for identification and recording is not lost.

> *The plan should include means for all those involved in any way with the change to identify resistance and discomfort of any kind.*

Questions for Identifying Resistance

- Who do we see as disagreeing with the project as we have presented it?
- Who is reluctant to participate in teams or devote time to what the project requires?
- Who has doubts about the training? Who is dissatisfied with the training? Who has doubts about their ability to apply the new knowledge? Who has doubts about their ability to maintain the new behaviors or routines?
- Who shows signs of having abandoned new practices and behaviors?
- Who is making organizational excuses for not doing what the project is asking of them?

DISTINGUISH BETWEEN WARNINGS AND RESISTANCE

In many cases, these are not resistances but warnings. For instance, inconsistencies between the proposed change and the unavailability of necessary tools and resources could give rise to such warnings. Or the continuity of policies that are contrary to the project and the intended changes.

These resistances serve as warnings, indicating the necessity for alignment. Their purpose is to guarantee that the project's objective is met. The project's design and execution rely on the identification and management of justified resistances. The greater the impact on the organization, the more crucial it is to address. In his book *Six Hats for Thinking*, Edward de Bono highlights the value of the Black Hat—or critical judgment and caution—in moments of analysis and decision making.

Listening to the motives and arguments of someone who opposes a change or someone who points out the downsides is extremely useful because:

- *They are right!* It makes sense to find aspects or components of the project that should be avoided. Why should the rationale, objective, and design of the project as we have envisioned it be the best one?
- *They are useful to us!* They show us ways to improve.

> *Listening genuinely and innocently to those who are resisting the project and talking openly and honestly with them is an excellent tool for a successful change plan.*

WHAT ABOUT THOSE WHO ARE SPONSORING THE CHANGE?

One should not forget the resistance that the sponsors, the promoters of the change in the eyes of the stakeholders, will unconsciously manifest. It is important to understand whether they form a solid coalition, whether they all have similar intentions and commitments to the project, and whether they form a solid front to face the undesirable consequences of the change. This is because they will have to set an example and start by changing some of their usual behaviors.

> *It is of interest to understand whether sponsors and executives form a strong coalition and whether they have similar intentions and commitments to each other.*

PREPARING THE RESISTANCE MANAGEMENT PLAN

Once it is understood how stakeholders will react to the expected changes, plans can be designed to explain, encourage, support, and sustain the change—see Chapter 20, "Checklist for the Planning."

Take stakeholder perspectives into account when designing plans to encourage, support, and sustain change.

At this stage of the change process, the aim is to address any identified resistance that cannot be prevented or mitigated by the communication, training, and support activities we have already discussed.

SOURCES OF INFORMATION

There are basically three sources of information to understand the resistances that appear or may appear—see Chapters 16, "Understanding Your Current Culture," and 18, "What Could You be Facing."

- *Project Culture and Stakeholder Impact.* This analysis should have triggered actions in the plans to create Awareness, Commitment, Knowledge, and Continuity through alignment, communication, training, coaching, and mentoring.
- *Continue to Listen to Stakeholders.* The dynamics of change require the consideration of the evolution of the project. Some resistance may not have been considered, may have arisen because of the actions implemented, or may have arisen from other stakeholders.
- *Listen to Employees*—at all levels and areas. Alignment activities, communication, coaching, training, technical meetings, informal meetings, or conversations are all good instances to hear the real opinion about the project and the changes. And even suggestions to decide on prevention and mitigation actions.

The key to addressing the issue of resistance is to identify its source. The most reliable source of information is the person or persons experiencing the resistance.

WHAT IF SOMEONE DOES NOT WANT TO PARTICIPATE

It is to be expected that not everyone will accept the project, particularly given the lack of clarity about the supposed benefit. In these circumstances, the win-win is for the project to meet their expectations of contribution.

The project must be implemented, but not wanting to participate in improvement teams, for example, is a position to respect. Isn't there any other way to have his/her collaboration? Or other ways to utilize their knowledge and experience than the one planned?

While collaboration and participation are necessary, they are a means to an end. The real objective is not to have 100% of the people involved, but to design, improve, and innovate—products, services, processes, and stakeholders' experiences.

NOTES

1 Elliot Jaques, Requisite Organization, *the CEO's Guide to Creative Structure & Leadership.* Cason Hall, 1992.
2 Define-Measure-Analyze-Improve-Control sequence used in the Six Sigma methodology.
3 Expression used in the Toyota Production System to identify the "place where things happen." Used in organizations implementing Lean.
4 Plan-Do-Check-Action sequence of problem solving. Used in organizations implementing Lean.
5 Key role in the implementation of Scrum, method for work cells applying Agile.

26 Recognition and Reinforcement

Raúl Molteni and Carlos Lucena

WHAT IS A REINFORCEMENT PLAN AND WHAT IS IT FOR?

A system for identifying, promoting, and recognizing examples that foster Awareness, Commitment, and Continuity.

IT IS DIFFERENTIATION, NOT JUST RECOGNITION

In Chapter 3, "The Psychology of Change," I mentioned the "2-6-2"; when faced with change, 20% of the people will join very quickly with enthusiasm and commitment; 20% will never join—at least not to the extent expected by the change proponents; and 60% will watch one another to take sides.

The key is to be nuanced with these groups. I prefer to highlight and recognize those who are making an effort and adding value to the project. But under certain circumstances, something has to happen to those who are not. "If nothing happens, why bother?" Penalizing actions, such as not to promote, have a prerequisite: having listened to, understood, and worked with them on their motives beforehand.

> *It requires distinguishing in some way those who adopt the new behaviors and acquire the habit of maintaining them from those who do not.*

CONSISTENT RECOGNITION

Recognition must be fully **aligned with the behaviors you are trying to promote with the project** and happens in real time. If you are trying to encourage collaboration and teamwork and give bonuses based on individual performance, everyone will eventually see your inconsistency. Such recognition may be deserved for reasons unrelated to the project, but it will not help change if it is seen as inconsistent and lacking in management commitment.

Recognizing one team for improving a process and others for individual work that is not consistent with the project requires great care in the design and delivery of communications.

DOI: 10.1201/9781003544807-32

MEANINGFUL RECOGNITION

Recognition must be meaningful, of perceived value to the recipient, visible, and ongoing. Some argue that there is no need for more recognition than something sober and symbolic because, after all, people are doing what they are supposed to be doing. Perhaps it is up to the HR function to identify the aspects to which employees are sensitive and to recognize them in the most encouraging and sincere way possible.

Whatever the element, the intended impact is not only for the person being recognized but also for the impact it could have on others and serve as a future incentive.

RECOGNIZING PROJECT LEADERS

It is the responsibility of the owner and the CEO to recognize the sponsors. It is the responsibility of the latter to recognize the facilitator's team.

ABOUT THE RECOGNITION PLAN

- Define how, when, and by whom recognition opportunities will be identified based on quick wins and other achievements, adoption of new behaviors, and use of monitoring data and learning instances.
- Develop a recognition plan reflected in a document clearly stating the goals and actions to be taken, why, and for what purpose. The specific activities of this plan are integrated with the other plans in the Global Plan.
- Secure resources from the organization.
- Establish deadlines for merit analysis based on the plan and project progress.
- Measure satisfaction resulting from recognition efforts.

QUICK WINS AS KEY REINFORCERS

Evidence shows that small wins are needed—both by those driving the change and by the rest of the workforce—to build Awareness, Commitment, and Continuity. It is true that projects have more important goals than a quick win. But remember, it's like celebrating a field goal. They show that the journey is worthwhile.

CONDITIONS FOR A QUICK WIN

- Short term.
- Consistency. With the values and practices associated with the project.
- No ambiguity; the indicators are reliable.
- Visibility. It is clearly identified, communicated, and believed.
- Credibility. Related to the effort behind the project.

Purpose to Be Pursued

- Evidence that the effort is worthwhile.
- Evidence that the project is on the right track.
- Evidence that the benefits are real.
- Evidence that it provides a basis for learning to improve and deploy.
- Evidence that it maintains Awareness, Commitment, and Continuity in the minds of the staff.

STRENGTHEN THE PROJECT BY CREATING A CONSISTENT ENVIRONMENT

Review and rewrite inconsistent policies, processes, procedures, and practices. Create connections and forget "divide and rule." Reinforce positive behaviors. Do not allow "nothing happens when nothing happens." These are examples of situations that must be eliminated to gain Credibility and Continuity.

In particular, recognition, performance evaluation and feedback, personnel selection, hierarchical promotions, rotation between functions, and the type of training and coaching are examples of aspects and programs whose consistency must be analyzed.

If you do not work to create an environment that is consistent with the change, the change will fit the environment:

- In some organizations and with projects and efforts aimed at improving customer satisfaction and experience, the cost of acquiring new customers is higher than the cost of retaining them. However, the salesperson's bonus remains tied to the sale. Unless the motive for that bonus is changed, the focus of the business system will remain on sales.
- An insurance company may want to have a more direct relationship with policyholders. However, if no one changes the relationship with the brokers, the brokers will continue to consider the policyholders their own and will not share the data with the insurer.
- The change may propose making the data related to machine downtime visible and reliable. However, if the operators' remuneration is related to the time of operation, the reliable declaration of the times will not appear.

It is the same system that contains the traps to slow down the change.

27 Metrics and KPIs

Raúl Molteni

THE RESULT OR THE OUTPUT

The output is the result of our actions, and the result is what we have achieved. The output of an activity includes communication about the change, content, media, and meeting the preestablished deadline. Despite our best efforts in design and delivery, the level of exposure or awareness we achieve may fall short of our desired outcome and be perceived as just another communication. What have we ultimately accomplished? Has it met our intended objectives, or have there been unintended consequences? While we can continue to rely on KPIs that focus on outputs, it is essential for the facilitation team and sponsors to develop KPIs that assess whether the desired outcomes are being achieved. Is it valid to track the number of individuals who attended the project presentation? We understand that it is, but did it have an impact on those who attended?

> *The output is the result of our actions, and the result is what we have achieved.*

ABOUT THE KPIS

It's no use having KPIs if you don't know what they tell you and how to use them. They may look nice in a PowerPoint and during a management presentation. But they do not add value. The function of KPIs is to provide information about what is really happening to identify opportunities for correction and improvement.

> *The KPIs themselves should be analyzed and reviewed to validate that they are "saying" what they are "supposed to be saying."*

We should understand whether the effective execution of a project leads the people involved to the new behaviors associated with its purpose. This is not a psychological question. It responds to the realm of evidence. Is something happening, and is it happening in the way defined as most appropriate? Can you measure the degree to

DOI: 10.1201/9781003544807-33

which people buy into a change, or is it an impossible mission? KPIs should show the response of how much and when behaviors and routines are changed, not just compliance with activities.

THE KEY: WHAT DO WE NEED TO KNOW?

From my perspective, the key question is: What do we need to know? If we wanted to know how convinced maintenance personnel were of the value of measuring MTBF (Mean Time To Failure) or MTTR (Mean Time to Repair), we could invent an indicator. However, it wouldn't take long for someone to question it.

However, we can rely on measurements to understand whether the situation is actually improving and evolving throughout the project and according to your plan. For indicators related to positioning for change, do we need indicators that tell us exactly how much it is? I do believe that precision is more important than accuracy. What we need to know is whether people are completely convinced of the need for change, or are they much more convinced than they were a month ago? Are people fully committed, or are they more committed than they were before they met with the coach a few times? If the answers are the latter, then we can identify and rely on an indicator.

For indicators related to positioning in the face of change, precision is more important than accuracy.

WHAT TO MEASURE

Determining what to measure requires much more analysis than you usually get. Not everything that can be measured is worth measuring. And sometimes we are held back by the difficulty of measuring what is worth measuring. The judgment applied to the analysis of indicators must have the statistical rigor it deserves. What we measure is data. We need information to manage the project. If we don't have quality data and do not interpret the data correctly, even the best measurement system will be useless.

Not everything that can be measured is worth measuring. And not everything that is worth measuring can be measured. But nothing justifies not having a balanced dashboard of the project and the change it brings.

A CHANGE MONITORING SYSTEM

A set of indicators to measure the impact and outcome of the Global Plan: (i) to validate the results of the plan and adjust when necessary and "on time"; (ii) to have information to support communication with data.

THE OBJECTIVE OF THE SYSTEM

Getting and analyzing the results has four purposes:

- Determine which actions were ineffective. Find the root cause and incorporate adjustments and corrections into the plan.
- Identify new opportunities. New scenarios or aspects that have not been considered and need to be worked on.
- Identify people who need special support and act accordingly.
- Identify successes and quick wins for recognition and celebration.

> *Once the indicators have been defined, a formal monitoring system must be designed to provide feedback and allow for adjustments to the plans.*

1. Index for Measuring Maturity in the Face of Change

These indicators show the evolution of individuals and groups throughout the project and change. For example, those aspects related to maturity levels[1] and to the evaluation of the sponsors.

I use questionnaires to assess the extent to which everyone has developed Awareness, Commitment, Knowledge and Competencies, and Continuity. It is important to ascertain how the individual in question perceives the level of leadership and commitment demonstrated by the organization. The results allow us to determine the appropriate course of action for that individual or group.

Each response to the following questionnaire can be evaluated on a scale of 1–5, with the monthly average and variation plotted. The results will demonstrate:

- Levels and trends for the degree of Awareness, Individual Commitment, Knowledge and Competencies, Continuity and Routines, and Organizational Commitment.
- Growth.
- Relationships between criteria. For example, to validate that the greater the Organizational Commitment, the greater the Individual Commitment.
- Similarities or differences between sectors' or functions' patterns.
- Information to identify root causes of behaviors and outline specific action plans—individual or group.

The facilitator team assesses sponsors, managers, middle management, and other project-related stakeholders, both individually and as a group, using the questionnaire. A Pareto diagram can assist in identifying key areas for improvement. Each question can be evaluated with 1 indicating "No," 2 indicating "Erratic," 3 indicating "Not always," 4 indicating "Continuously," and 5 indicating "Role Model" (Figures 27.1 and 27.2).

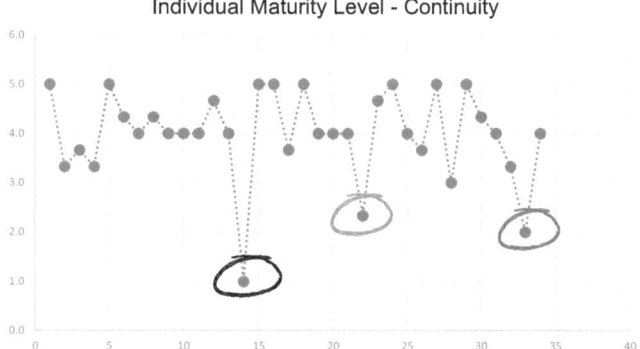

Individual Maturity Level - Continuity

Results suggest special control of three persons (1 PO from Commercial
and two from Finance -TI y PO).

FIGURE 27.1 Individual maturity level results regarding continuity for individuals in a certain foundry sector. The results for each person of the sector are shown in the vertical axis while the names are shown in the horizontal axis. It shows three individuals for whom additional support activities—further training and coaching—were planned due to the low results.

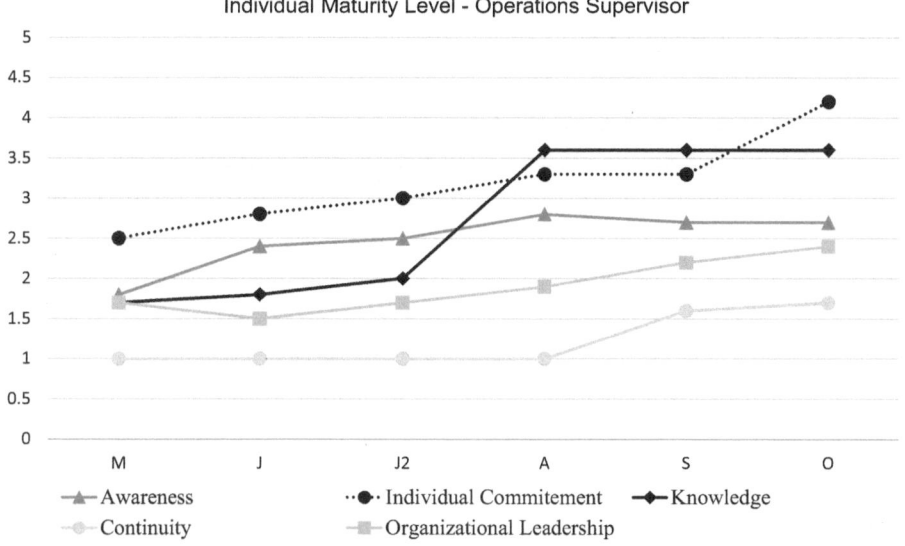

Individual Maturity Level - Operations Supervisor

FIGURE 27.2 Monthly evolution of the Individual Maturity Level of a Supervisor from Operations. Shows the improvement of the five factors: Awareness, Individual Commitment, Knowledge, Continuity, and Organizational Leadership.

PHASE OF MATURITY
AWARENESS
1. Understands and accepts the need for change.
2. Understands and accepts the benefits and opportunity for change.
3. Asks questions and inquires about his/her concerns about the change.
4. Gives public signs of acceptance of the project.
5. Delivers messages that are consistent with the vision for change. Avoids messages and actions that contradict the intended change.
INDIVIDUAL COMMITMENT
1. Shows willingness to share his/her doubts and the risks he/she sees for himself/herself.
2. Asks and inquires about his/her doubts about his/her future.
3. Wants to be part of the change through active participation. Shares spaces and moments of change with others.
4. Demonstrates behaviors consistent with the expected change.
5. Demonstrates a desire to understand more and be able to do more. Delivers what is asked and within the expected time frame.
6. Accepts being part of initial and pilot projects. Contributes suggestions and ideas.
7. Talks about what he/she does. Promotes the change.
KNOWLEDGE AND SKILLS
1. Accepts training.
2. Seeks to implement changes in his/her practices and behaviors.
3. Seeks and accepts support to implement changes in practices and behaviors.
4. Suggests changes in policies, processes, and procedures that inhibit him/her to change.
5. Talks about what he/she has learnt.
6. Demonstrates a willingness to learn more about change and what to expect from it.
CONTINUITY AND ROUTINES
1. Continually demonstrates new behaviors and practices in daily activities.
2. When finds barriers, he/she highlights them and works with other people together to eliminate them.
3. Seeks and accepts support to implement changes in practices and behaviors.
4. Accepts new training activities.
5. Suggests changes in policies, processes, and procedures that inhibit change—in general and for himself/herself.
6. Argues, manages, and makes decisions based on relevant information.
7. Clearly communicates his/her own thoughts and suggestions for the development of change.

PHASE OF MATURITY
LEADERSHIP AND ORGANIZATIONAL COMMITMENT
1. The intended change is directly linked to the strategic goals of the organization. It is communicated in this way.
2. The vision of the change and its rationale, sense of opportunity, benefits, and risks are continuously communicated.
3. Recognition is provided on an ongoing basis, differentiating those who join from those who do not.
4. Small changes are fostered and celebrated to encourage those who do and to invite those who do not to participate.
5. Individual efforts are reinforced with subprojects to align and simplify policies, processes, and procedures.
6. The evolution of the change is continuously evaluated using indicators
7. Myths and stories are recreated to reinforce the why and the evolution of the change by appealing to logic and emotion.

2.   Level of Adherence to Critical Behaviors—Systematic Assessment

These are indicators of behavioral compliance. They measure the degree of compliance with behaviors or routines that must be performed systematically. The steps to follow are:

1.   Determine the behaviors or routines to be assessed and the conditions that must be met to be considered satisfied.
2.   Determine the situations and times when they should be encountered.

An Example

(i) Customary behavior: Decisions are delayed at the weekly management meeting; the meeting ends without a decision being made, and the current situation is maintained as a result; (ii) Expected behavior: At the scheduled meeting, a decision is made to favor one of the options presented or a new option developed at the meeting; (iii) To be observed in: All weekly meetings; and (iv) Who observes and records: Whoever is designated as secretary for the day. See Figure 27.3 for better clarity.

3.   Level of Adherence to Critical Behaviors—Incidents

These indicators are used to verify the extent to which people learn new behaviors and apply the concepts of the change in situations that are not routine and planned.

Behavior Compliance Matrix

Habit to modify	Activity	Desired habit	Y	N	%	M1°	M2°	J1°	J2°	J1°	J2°	A1°	A2°	S1°	S2°	O1°
There is no widespread habit of using indicators or visual management boards	Scoreboards in North plant	The information on the boards is complete and up to date	4	2	67	Y		Y		N		Y		N		Y
There is no widespread habit of using indicators or visual management boards	Scoreboards in Ferva plant	The reasons for the main deviations on the board are identified.	3	3	50	N		N		N		Y		Y		Y
Work is done in silos and each office is an island. There is no collaboration between areas	Team meetings in North and East plants	The inter-plant meeting was held at the time and with attendance as per the call.	5	0	100	Y		Y		Y		Y		Y		
Decisions are delayed and the status quo ends up winning.	Team meetings in North and East plants	The actions committed to on the established dates were monitored.	4	1	80	Y		Y		Y		N		Y		

FIGURE 27.3   The Behavior Compliance Matrix includes the compliance and noncompliance of various employee behaviors in a given area. The habit to modify—the current—the activity being executed, and the desired habit—what should be observed after training— are observed, resulting in the nonconformance index that drives improvement actions.

Each incident is a situation that demonstrates a behavior that is consistent or inconsistent with those required by the project. It is weighted according to the impact, positive or negative, with respect to what would be expected and the number of people exposed to the behavior. It is a way to identify and recognize small quick wins. I use the following criteria:

- *Positive and consistent impact on the project.* It is a positive example—role model—for the project (5: highly consistent behavior and visible to a large group of people; 3: highly consistent behavior and visible to a limited number of people, 1: positive behavior, but without significant visibility).
- *Negative impact and consistent with the project.* (-5: highly inconsistent behavior and visible to a large group of people; 3: highly inconsistent behavior and visible to a limited number of people; 1: inconsistent behavior, but no significant visibility).

Examples of Behaviors with a Positive Impact on the Project

Supervisors posted ideas for improvement on the department bulletin board. The supervisor actively participated in creating the VSM. The manager and supervisor actively participated in retrospective meetings. The area manager publicly thanked everyone involved in solving the input problem. The area supervisor asked for help assembling the mix/piece indicator. A foundry operator pointed out an accident hazard caused by a missing warning sign. The area supervisor immediately installed the warning sign.

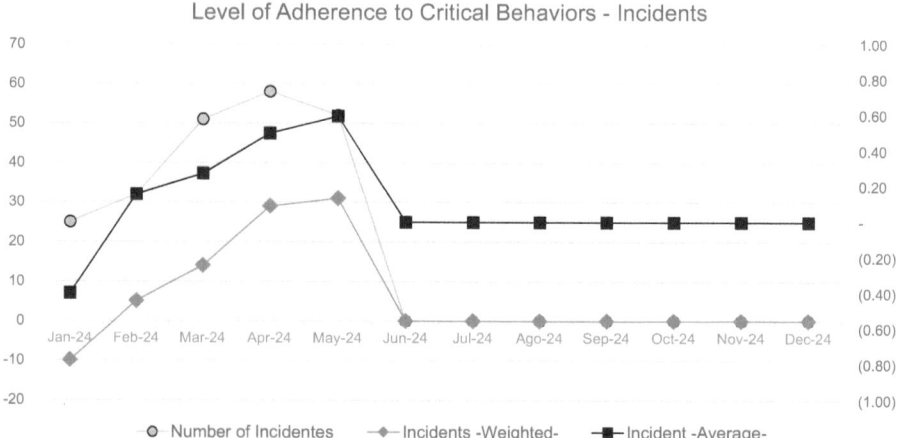

FIGURE 27.4   Monthly evolution of positive and negative incidents—what someone does or says that is consistent or opposed to the desired by the change. The chart shows a tendency to identify a more significant number of incidents in line with what was expected by the project, as well as the persistence of incidents as opposed to what was expected.

EXAMPLES OF BEHAVIORS WITH A NEGATIVE OUTCOME ON THE PROJECT

The area manager had not yet started the performance analysis meetings and was 17 days late. The area manager did not leave his office to attend the training of his supervisors. A foundry operator reported an accident hazard due to a missing warning sign, and his supervisor did not take action (see Figure 27.4).

4.   Technical Indicators

These are the indicators associated with the technical side of the project. For example, quick wins results, the results of improving a process in terms of quality, time, and cost, the number of observations of the internal ISO 9001 audit, or the burndown chart in Agile cells.

HOW TO MEASURE THEM

On the basis of the result of the best practices related to the technical methodology of the project.

5.   Examples of Other Indicators Related to the Maturity Phases

This is an additional list of indicators that could be useful to track the evolution of the project and its changes, and the degree to which new behaviors are being accepted and adopted by the staff.

TABLE 27.1
Results of each phase by sector

Phase	Sector 2	Sector 3	Sector 4	Sector 5	Sector 6
Awareness	3,45	2,41	3,41	3,55	4,44
Commitment	3,78	2,60	2,73	3,56	4,14
Knowledge	3,00	2,53	2,73	3,35	4,03
Continuity	2,04	1,65	1,43	3,22	4,00
Organizational Commitment	1,00	1,19	1,42	2,22	4,00

AWARENESS

- Number of people who understand the change and why it is needed.
- The tone of comments (opposition, neutral, support ...) weighted by the source's credibility.
- Type and number of questions, doubts, and suggestions in internal forums.

INDIVIDUAL COMMITMENT

- Number of people directly and actively involved in the change.
- Degree of rotation among group members.
- Number and rating of sponsors.

KNOWLEDGE AND COMPETENCIES

- Number of people prepared for change with validated readiness.
- The gap between current and required knowledge and skills.
- Type and number of questions, doubts, and suggestions in internal forums about required competencies and knowledge.

CONTINUITY AND ROUTINES

- Results of audits of adherence to new behaviors and habits.
- Results of the quick wins.
- Results achieved with the changes.
- Number of people and areas involved in the change.

Leadership and Organizational Commitment

- Employee perceptions of Leadership Commitment.
- Perception of recognition.
- Number of processes, procedures, systems, and mechanisms updated to align them to the change.

NOTE

1 Awareness, Commitment, Knowledge and Competencies, and Continuity phases—based on Prosci's ADKAR.

Part VII

Implementing and Sustaining the Change

This section outlines the third phase of the roadmap. This phase is focused on implementing the strategy and plans, based on the results of the previous phase. The objective of this phase is to gain control of the plans, facilitate the learning process, and implement feedback and improvements to the strategy and plans.

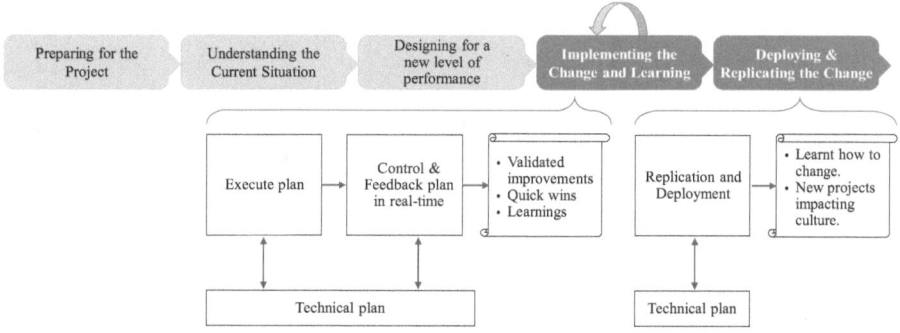

FIGURE 0.7 Steps of the Implementing and Sustaining the Change phase: Execute the Plan, control, provide feedback in real time, and learn. These steps are discussed in Chapter 28. It also shows the Phase of Deploying and Replicating the Change, which is discussed in Chapter 29.

DOI: 10.1201/9781003544807-34

28 Implementing the Change and Learning

Raúl Molteni

ABOUT EXECUTION

It is the practical application of project management methodologies, such as those outlined by PMI. The strategy and plans developed in the previous stage serve as the foundation for this stage.

ABOUT CONTROL

At this stage, it is also necessary to install the Project Control Dashboard, and the indicators suggested in Chapter 27, "Metrics and KPIs."

ABOUT FEEDBACK

The strategy and plan must be based on feedback as well as the results—metrics and KPIs. The suggestions in Chapter 20, "Checklist for the Planning," serve as inspiration to understand the options available based on the results and learning.

ABOUT LEARNING

Learning is based on the concepts discussed in Chapter 5, "About Continuous Learning." Learning objectives at this stage are as follows:

- Identify, review, and redefine the concepts that have led to a design decision—a process, product, service, project, strategy, or plan.
- Identify new decision and action options for situations like those experienced.
- Identify decision and action options for situations different from those experienced.

WHAT TO LEARN

Learning throughout the project, while learning to learn, involves three distinct focuses.

DOI: 10.1201/9781003544807-35

- *Skills*. Usually, to perform tasks that have been redesigned or are new.
- *Cognitive Process*. Related to the ability to analyze information and make decisions.
- *Social*. Related to interaction, communication, and changes in behavioral patterns that are part of the change.

LEARNING WHILE ACHIEVING EXPECTED RESULTS

Learning is not a necessary skill to be applied only when results are not achieved. It is learning from what works and what does not, under what conditions, and in what circumstances. The key is to learn from what happens to repeat or improve in the future. Let's analyze these situations:

- *Results were positive and were achieved by applying the plan and practices*. Was it this plan and these practices that led to the positive results? What are the opportunities for improvement for the next project or practice?
- *Results were positive, but the plan and practices were not implemented—at least not completely*. Where did the results come from? What variables "helped"? What variables were not considered? What should be kept, changed, or added in similar situations?
- *Results were not positive, and the plan and practices were used*. Were there variables that were not considered? Were the causes poorly assessed? Were the solutions misaligned with the causes? Were the solutions ineffective?
- *Results were not positive and the plan and practices were not implemented—at least not fully*. Why was the plan not followed? What would be the next step?

Each of these responses should lead not only to corrective action but also to other types of responses. Perhaps more importantly, why were we wrong in interpreting the situation or generating solutions? It is quicker to try to correct the situation. But this does not necessarily allow for learning. There may be insufficient understanding of what to think and do differently on new occasions.

> *Learning is not a necessary skill to be brought into play only when results are not achieved. It is learning from what works and what doesn't.*

SECOND-ORDER LEARNING

Second-order learning is an important part of the facilitator team's role. It involves understanding what reasoning, belief, or knowledge led to defining an action that had not worked. It refers to in-depth knowledge of the successful action.

Under which beliefs, with which concepts, and with which knowledge, techniques, resources, and competencies should we design, operate, and control a new project?

Second-order learning should be a fundamental weapon in the arsenal of an organization's board and executives. To undertake meaningful projects, mainly those that involve "changing the culture," second-order learning must begin with the owner, the board, the CEO, or the head of the unit. The questions to be answered should be:

- What beliefs, knowledge, paradigms, and experiences have led us to adopt and approach previous projects with the results we had?
- What beliefs, knowledge, paradigms, and experiences lead us to approach this new project in the way we are? Are they the same?

Owners, board members, and CEOs must generate reflection and become observers of their own actions, thoughts, and decisions.

LEARNING WORKSHOPS AND MEETINGS

Learning workshops should include activities that the team of facilitators can coordinate, with representatives of all stakeholders participating. They should be systematic throughout the project and at its end.

The following steps should be taken to conduct a learning workshop:

- Identify the key aspects or variables of analysis according to the project. What aspects should be reflected on?
- Collect data related to the practices and plans and the respective results.
- Prepare those who are to participate, making the objective clear.
- Complete, together, the Learning Matrix. Analyze (Figure 28.1).
- Improve the project plans, if necessary—it is in most cases.
- Think which of the principles, theories, and assumptions these plans were based on could be questioned in light of the analysis made.

QUESTIONS TO GUIDE REFLECTION AND LEARNING

Here is a checklist to facilitate learning meetings:

- How clear were we, and are we now, about the intended change in scope and goals?
- How do we feel about the project? Have we made progress on the aspects we intended? Is it working or not according to the objectives? What does the data tell us?
- What is working well, why, and under what conditions?

- What is not working, how do we want it to work, why and what for, and under what conditions?
- Is the technical development of the project consistent with the objectives?
- How are the results being achieved compared to the defined goals?
- How do we perceive the relationship and functioning of our sponsors' coalition?
- How do we perceive the relationship and functioning of our facilitator team?
- How was—and is—the project prepared and supported by the sponsors?
- How has the support and involvement of senior and middle management been?
- How do we perceive the relationship and functioning of the other groups involved in the project?
- How do the sponsors and sponsors' coalition evaluate the results?
- How do leaders evaluate the results?
- How do the stakeholders evaluate the results?
- How was the technical preparation and learning of the facilitator team?
- How do the executives prepare for and support the project?
- How successful is the communication and ongoing involvement of all stakeholders?
- How effective is the planning and ongoing management of the project?
- How does the planned and ongoing change management work?
- How does the planned and ongoing risk management work?
- Are the resources allocated—financial, technical, and time—sufficient and appropriate?
- How well did we address stakeholder needs and expectations?
- How well did we prepare for this learning event?

ABOUT THE SECOND-ORDER LEARNING

- What were we convinced of, and the results proved us to be wrong?
- What beliefs did we have, and what principles did we use that we should question considering all the answers above?
- What beliefs and principles should we embrace in further projects?

Owners, board members, CEO, and unit managers must become observers of their own actions, thoughts, and decisions.

THE PRIORITIZATION MATRIX

The following Matrix helps to focus on the more important learnings. It allows different criteria depending on the type and objectives of the project (Figure 28.1). We propose to consider:

- *Impact on Culture.* Evaluated on a scale of 1–5, with 5 being very satisfactory.

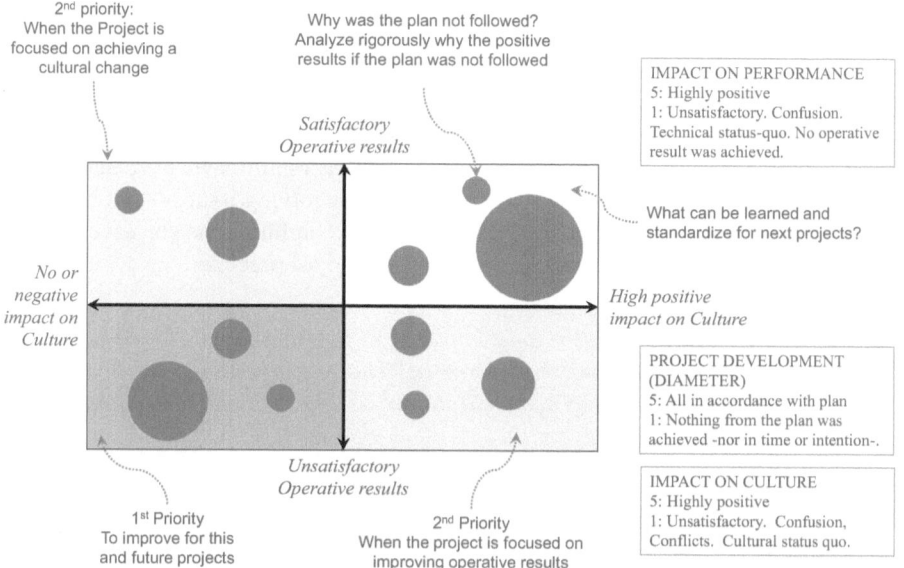

FIGURE 28.1 Learning Prioritization Matrix. It shows the three criteria for analyzing project results: operational performance, impact on culture, and the degree to which strategy and plans were achieved. Moreover, by assessing each factor of the project and graphing on the matrix, the priorities for analysis and learning become clearer.

- *Performance Impact.* Evaluated on a scale of 1–5, with 5 being very satisfactory.
- *Project development.* Evaluated on a scale of 1–5, with 5 in line with the project's strategy and plan.

LEARNING CONCLUSIONS

- *Outcomes*: What are the real benefits of this project in terms of operational, cultural, or learning outcomes?
- *Ineffective Actions*: What are the root causes? What are the plans and actions that are not effective in achieving our goals?
- *Opportunities*: What are our opportunities to go "beyond," to improve, to deploy, or to initiate new projects based on what we have learned from the project?
- *Celebrations*: What should we celebrate, with whom, why, and what for?
- *Significant Lessons*: What are the most important Second-Order Learnings?

ABOUT RECOGNITION

It is important to recognize those who are and have been involved in the project's development. However, the plans should determine how, who, when, and why teams and/or individuals in the organization will be recognized. What success can we celebrate? What behaviors can we reinforce? Who deserves recognition for their efforts and collaboration?

As with communication, recognition is never sufficient. Therefore, it is recommended that acknowledgment be provided even if the results and changes seem minimal in comparison to the desired outcomes. The recognition we are discussing is not only intended to acknowledge the individual for resolving issues—that he or she may not have created. The purpose of recognition is to highlight the correct course of action and demonstrate the desired behavior as a standard practice.

> *Recognition is not intended to acknowledge the individual for resolving issues—that he/she may not have created. The purpose is to highlight the correct course of action and demonstrate the desired behavior as a standard practice.*

SUGGESTIONS

- Identify and consider what is meaningful to the individual or group being recognized.
- Define the structure of the recognition program, including the objectives, the criteria, and the metrics for evaluating the impact of these actions.
- Obtain information from within the organization regarding existing institutional practices for recognition, as well as those that have been unsuccessful in the past and the reasons for this.
- Identify by analyzing satisfaction surveys or information collected in training and coaching conversations—among other sources—the impact and success of recognitions given.
- Determine whether there is a perception of deserved and omitted recognition.

29 Deploying and Replicating the Change

Raúl Molteni

SUSTAINING CHANGE

Sustained change implies that a transformation has already occurred within the organization and has permeated its culture, at least primarily among those involved. If the change has been implemented in accordance with best practices, the organization's ability to maintain it over time will depend mainly on the discipline with which it exercises established routines. No less important will be the continued alignment with the principles and values that gave rise to the project and the change.

Some of the "virtues" or characteristics that should prevail are:

1. *The management's conviction*—directors, CEO, sponsors' coalition, executives—that the principles, values, and objectives that gave rise to the project are still valid for the organization.
2. Maintenance of *activities to identify the obstacles, concerns, and difficulties* of the employees, as was done during the project.
3. Maintaining a *training process—Knowledge and Competencies*—and a Knowledge Management that protects improvements from staff turnover and loss of knowledge—see Chapter 30, "Knowledge Management."
4. Maintaining a *willingness to review policies, processes, and procedures* that prove to be barriers to achieving sustainability.
5. Maintaining a *system of data and indicators* to ensure that the practices that should be adopted are being maintained and that the results that should be achieved are being achieved.
6. *An improvement system* that, beyond the results achieved by the project, maintains the concern—and the methods—to interpret the stakeholders and introduce changes that achieve the evolution of practices and results.
7. The *discipline to maintain activities*, processes, procedures, and controls under this improvement system.

DOI: 10.1201/9781003544807-36

8. *Use of standards*—such as ISO 9001, 14001, and those corresponding to the sector, type of process, and type of project—and a system of internal audits.
9. Responding to the management and with a real interest in improvement—*regardless of whether there will be an external auditor* to whom they will have to respond.
10. Synchronization and alignment with and between other projects that may arise.

> *What happens when others join? What happens if new players override something?*

The same goes for the methodologies, processes, and practices that were key to the implementation process and whose sources must be kept under control:

- The desire to learn.
- Risk management.
- Managing unpredictable events.
- Resistance management.
- Alignment between affected systems.

SUSTAINABILITY HAS ALREADY BEGUN

It is important to reiterate that the project process is not always linear, as may be interpreted from this and other books. Cycles of review, learning, and feedback mark stages and "restarts." There is no such thing as "time for working on the project's sustainability." Sustainability should be built throughout the project and should have been considered in each step and tool previously applied.

> *There is no such thing as "time for working on the project's sustainability." Sustainability has to be built throughout the process.*

A CLEAR INDICATOR OF SUCCESS

A key indicator of the success of the project in terms of its impact on the organizational culture will be the extent to which personnel at all levels and across all sectors are actively engaged in maintaining the changes that have been implemented.

SUSTAINING IS NOT MAINTAINING

Sustaining change is not simply maintaining it. It implies keeping a questioning attitude toward the variables of success, challenging them, and allowing oneself to recreate approaches and instruments. We might assume that sustainment requires less effort than implementation. However, the culture and talents of each organization may be more accustomed to founding or building from scratch, while others are more accustomed to standardization, discipline, and sustained repetition. These aspects also count.

ENEMIES OF SUSTAINABILITY

In *The Dynamics of Planned Change*,[1] it is argued that one of the most pressing problems of sustainability is the lack of diffusion among neighboring systems or subparts of the affected process. As previously discussed, this concept of integration and coordination between functions and projects is fundamental to the success of any business.

Another enemy to mention is Inertia. Not only the one that inhibits leaving old habits behind, but also the one that—like a spring—pulls us back to take refuge in a familiar status quo. We must challenge these physical forces with driving forces that neutralize them. The book mentioned earlier speaks of enduring "certain changes simply because the progressive movement of the system is a more powerful force than any of its incipient retrograde tendencies."

Robert Quinn[2] emphasizes that comprehensive change—both personal and corporate—is ultimately a spiritual process. Let us consider the implications of a process of adopting and sustaining a change that we have previously experienced. We will confirm that it is reasonable to recognize that this dimension is present. Examples of this include setbacks experienced by individuals who have abandoned a harmful habit or experienced a diet-related relapse.

REPLICATION AND DEPLOYMENT AS ELEMENTS OF SUSTAINABILITY

Replication refers to applying a project or solution to other units, functions, processes, products, or services. I refer to deployment as applying the project or solutions to the complete unit, function, process, product, or service. The specific approach ultimately depends on the chosen strategy.

A project that delays deployment or replication raises concerns about its future viability. However, the learning gained determines when and how it must be made. An early deployment could leave an unconsolidated change that defeats all subsequent efforts when reversed in its original area.

ABOUT THE DEPLOYMENT PLAN

Once the initial project—pilot or otherwise—is complete, the deployment should be planned. The characteristics of the culture at the time and the technical requirements will serve as a guide. However, ensure the project has been consolidated beforehand in its original scope. Deployment can be:

- *Horizontal.* From one function to another or from one sector to another.

- *Vertical*. One or all of the functions, but with the project being driven first by the managers, then by supervisors, and then by other operational levels.
- *By priority*. According to priorities determined by a previously selected criterion. Examples include customer impact, work environment impact, or cost impact.

NOTES

1 Ronald Lippit, Jeanne Watson, and Bruce Westley, *The Dynamics of Planned Change*. Amorrortu Editores, 1958.
2 Robert Quinn, *Wisdom for Change*. Prentice Hall, 2015.

30 Knowledge Management

Raúl Molteni and Carlos Lucena

ABOUT KNOWLEDGE MANAGEMENT

We can define knowledge management as a system that enables, through specific processes, to:

- Keep the knowledge gained from work and learning within the organization and not be affected by employee turnover.
- Make the knowledge available at the right time and place to develop similar projects, as well as to maintain the changes resulting from the project and manage the day-to-day operations.
- Innovate by integrating data and information from different sources and disciplines.

DIFFERENCE BETWEEN INFORMATION AND KNOWLEDGE

The information management process aims to systematically capture, select, organize, filter, present, integrate, and use it by the members of the organization to collaboratively leverage their intellectual capital and improve the organization's capabilities. When this information enables the creation of new approaches, it becomes knowledge. And when this creation is more related to strategy than technology and content, it is the starting point for a knowledge-based organization.

> *Knowledge management has little to do with machines and technology. The process is 80% social and 20% infrastructure.*

Information refers to content—data, descriptions, instructions—maintained in records such as spreadsheets, reports, and presentations. It enables us to know what is going on. Integrating new information with previous experience and knowledge allows us to gain new knowledge; it refers to the use, construction, and creation that allows us to interpret, understand, and contextualize this data. Understanding what is happening is not necessarily enough to conceptualize the situation in its context, relate it to other

DOI: 10.1201/9781003544807-37

events and data, intertwine it with other situations, causes, and actions, and ultimately solve more complex situations.

OBSTACLES AND RESISTANCE

Knowledge management faces more difficulties than technological ones regarding where and how to store information. For example:

- *"Knowledge is power."* As a result, there will be people who do not want to share what they have learned.
- *Not understanding the difference between data, information, and knowledge* and believing that if you have data on a server, you have knowledge.
- *Insufficient or inconsistent recognition and reinforcement.* For example, promoting those who do not share information or not giving enough credit to those who do.
- *The cost and difficulty* of finding adequate means to store and share knowledge.
- *Mistrust of the sources of knowledge,* such as, denial of "what wasn't invented here."
- *Concerns about data confidentiality.*

These are the difficulties that the sponsors' coalition and the team of facilitators will have to work on, using the same mechanisms we have seen to approach the project.

KNOWLEDGE MANAGEMENT PROCESS

The knowledge management process can be illustrated as follows:

IDENTIFICATION

It is the identification of the available knowledge, using techniques and tools, especially those generated during the project's development. It includes who has it, where it is registered, and its potential.

ACQUISITION

Knowledge may be "inside the organization," but acquisition refers to "capturing" it, materializing it, and recording it in a comprehensible and usable way. Learning meetings and workshops are an excellent source of acquisition, as are the people involved in the various analyses of the technical, social, and management aspects of the project itself.

Knowledge may already be available, or there may be a need to acquire specific knowledge leading to internal or external sources—this may have occurred throughout the project, for example, in benchmarking studies.

DISSEMINATION AND RETENTION

It is about making knowledge available to the people who might need it. Formal spaces for interaction and exchange between these people formally preestablished, the installation of specific software that ensures the updating and the absence of duplication, the incorporation in the project management methodology, the incorporation in the documentation of processes, procedures, and instructions, and in the training and leveling programs are some of the means available.

USE

It is necessary that knowledge is used to guarantee an adequate response of the organization to the internal and external situations to be faced. The use of knowledge in new projects, in problem-solving, and in daily operations becomes essential for the organization to be able to respond adequately and agilely to the continuous changes in the environment and not to "reinvent the wheel" with every new circumstance.

ASSESSMENT

In this phase, the entire knowledge management system is evaluated and improved. Several questions guide the analysis:

- Is it of use?
- Is it visible to those who need it?
- Does it enable us to understand what we need to understand?
- Does it allow us to take advantage of what we learn?
- Are we left without knowledge when someone leaves the organization?
- Do we have the knowledge we need for our current operations?
- And in terms of strategy, do we have the knowledge we will need in the future?

ALL ABOUT THE PROJECT IS ONLY A PART OF KNOWLEDGE MANAGEMENT OBJECTIVES

Knowledge management works by facilitating the creation of previously studied skills and routines that help people perform tasks without thinking about how and when to do them. As if we act mechanically when faced with situations similar to others from which we have learned, it frees our mind to devote it to aspects such as assessing situations and making decisions. The application of a methodology guides us to:

- Acquire knowledge and competencies because of the design, development, and sustainability of the project, based on data and successive learning.
- Maintain what has been learned during the project and sustain changes in the personnel operating it, whether current or potential future personnel due to rotations.

- Facilitate the use of existing knowledge through socialization, combination, externalization, and internalization processes in the sustainability of the project, in new projects, and in new undertakings of the organization.

PROJECT STORYBOARD AS PART OF KNOWLEDGE MANAGEMENT

All changes, whether in policies, processes, procedures, records, controls, and responsibilities, among others, should have been updated specifically, precisely, and simply in documents—virtual or physical—and be accessible to all those who, for any reason or any means, should know about them.

Moreover, the storyboard, or the memory of a project, is a useful tool for organizing and transferring knowledge, even if it responds to what has been done.

The key is not the project's storyboard or to have everything about the project written or to have it neat. It is to share what is learned so it can generate more value in the future.

KNOWLEDGE MANAGEMENT AND AI

Looking deeper, knowledge management is currently applied in organizations and has much more to do with data reservoirs than knowledge. AI will help to find the necessary information and interconnect it for the creation of new information, technology, and other knowledge.

Innovate by integrating data and information from different sources and disciplines.

Unlike "believing" that every problem can be solved with a unique problem-solving methodology—be it PDCA, McKinsey 7 steps, Six Sigma, Scrum—or a design methodology—such as Design Thinking—AI will most likely help us determine and use a better methodology for the particular problem we are addressing: a kind of methodological shortcut.

This panorama shows us, once again, the need to think about the future with interfunctional teams; strategic thinking seems to be something that will tend to gain relative weight. Why will we need AI? Why and how will we be using it?

Part VIII

And Now What

This section describes the ways of sustaining the continuity of projects in the organization.

DOI: 10.1201/9781003544807-38

Part VIII

31 Renewing Motivation and Continuity

Raúl Molteni

FIRST, TAKE A MOMENT TO BREATHE AND CELEBRATE

You could have encountered difficulties of all kinds: operational, commercial, economic, and social. You could have overcome them, though not all of them. There could have been some heartache and pain. Had this not been the case, the project would not have been as impactful for the organization. There will still be actions to implement or complete. There will also be data to analyze to validate the solution and sustainability fully.

But if the project, as I experienced it, had challenges, results, high satisfaction, and difficult moments and consolidated a team with great potential, capacity, and motivation, I invite you to celebrate. Include the owners, the directors, the CEO, the executives, the sponsors' coalition, the team of facilitators, and the rest of the organization. It is well deserved. The learning must be very valuable. Most surely, the organization has taken a step forward in the search for quality excellence and has evolved.

Moreover, a project seems to end as if we have reached a goal. In reality, a project that leads to a transformation of the organization—in terms of strategic, operational, or social positioning—is a never-ending process. It is a "never-ending improvement" (see Figure 31.1).

IT IS TIME TO EXPAND THE SCOPE OF OPERATIONS

As part of this process, we must now scale and deploy the knowledge gained. It is important to continue and deepen the objectives, methodologies, knowledge, and competencies gained from the project:

- As was said, replicate the experience in other functions, areas, processes, and units that were outside the original scope of the project.
- Apply the experience gained to projects that complement the one previously undertaken or that seek to capitalize on opportunities that have arisen as a result of it.
- Apply the experience to projects that seek to capitalize on other and new opportunities.

DOI: 10.1201/9781003544807-39

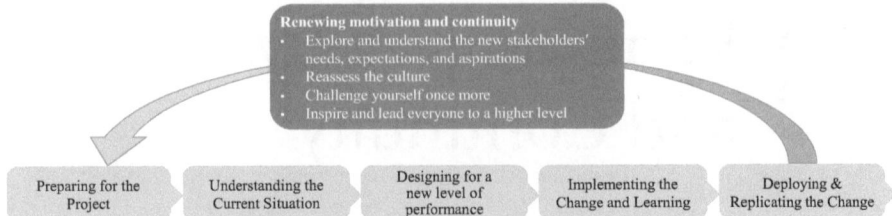

FIGURE 31.1   Closing the improvement loop and going back to prepare and develop another project.

It is also time to extend invitations, provide training, and bring on additional personnel at all levels and functions. To achieve critical mass and integrate concepts, methodologies, and techniques useful for improvement projects and applicable to day-to-day operations.

CONCEPTS TO BE REINFORCED

In conclusion, I would like to draw your attention to concepts introduced throughout the chapters, which I consider fundamental for future development.

- *Continuous improvement, innovation, and creativity.* More than ever. The shelf life of a solution is getting shorter and shorter.
- *Continuous learning.* Learning through evidence, through data, through methodology. Experimenting to avoid blindly following disinformation and false experiences.
- *Constant validation of knowledge.* Our technology references coin a phrase: "Those who say they know what is coming in the future in terms of digitalization and technology, in fact, do not know." AI threatens to kill us—humans—and another day, it seems to kill itself.
- *Technology at the service of business.* As Willy Vandenbrande[1] put it, "It's not a question of being against technology. The risk is that we see technology as a goal and want to implement it whether it's interesting or not." It is about technology at the service of stakeholders and strategic goals. Technology is the key; it helps to integrate visions and disciplines, to facilitate solutions for improvement and transformation, and to better relate to each other—all stakeholders. The obsession with technology because of technology itself, misinformation, and concern for the confidentiality of data and information require care in the selection and implementation.
- *Leveraging AI,* machine learning, IoT, AR, VR, and all the other technologies and combinations of technologies being offered to us.

ORGANIZATIONS NEED TO ACQUIRE OR IMPROVE SOME COMPETENCIES:

- *Define and implement a purpose* with which employees can identify. I am not talking about a well-defined political statement; I am talking about a real conviction to consider and respond to social needs, such as alignment with one or more of the United Nations' Sustainable Development Goals (SDGs).[2]
- *Define and install a strategy* for the desired change that deals with the uncertainty.
- *Understand employees*, their archetypes, their needs, where and why they get joy, and where and why they get pain along their work journey.
- *Understand and improve the employee experience* and how it aligns with strategy.
- *Find out how you can empower* people at all levels and let them take the initiative to respond to all stakeholders. And, of course, how to better use all *data analysis techniques.*
- *Measure and track your evolution.* Have a scoreboard. Since transformation requires a different mindset, understanding the key metrics to install that would indicate if the path drives real and effective transformation is key. What set of metrics can capture how people are moving through the stages of change, how sustainable the implementation of new habits and behaviors is, and how quickly these metrics are improving? Remember, you're not just doing this to change culture. So, technical metrics should be added to your scoreboard.

BEGIN ALL OVER AGAIN

As in the PDCA, embrace new opportunities and challenges with a fresh perspective:

- Explore and understand the new needs, expectations, and aspirations of your customers, employees, suppliers, and the society your business impacts.
- Reassess the culture, understand what changed, and discover what new beliefs and behaviors will be necessary for future success.
- Challenge yourself once more.
- Inspire and lead everyone to a higher level.
- And have fun.

And don't forget to integrate the pursuit of quality, excellence, and continuous evolution into all aspects of your life and work as a foundation of everything you do.

ABOUT THE FUTURE

Two sentences at the end to reflect the adventures that led us to write this book:

The future is built every day; it is reformulated every day.

"The people who are crazy enough to think they can change the world are the ones who do." Steve Jobs[3]

NOTES

1 Willy Vandenbrande, IAQUIS Academy.
2 United Nations, *17 Goals to Transform Our World*. United Nations, 2023. www.un.org/sustainabledevelopment.
3 Walter Isaacson, *Elon Musk*. Debate, 2024. From Apple's advertising campaign in 1997, "Think Different".

Part IX

Testimonies and Case Studies

This section describes one testimonial of how culture impacts projects aiming at transformations based on process management improvements and digitalization. It also addresses two case studies that show successful transformations. All these show the importance of having and considering the culture as a key to sustainable results.

DOI: 10.1201/9781003544807-40

32 Inhibitors to Project Success Deeply Tied to Cultural Dynamics

Nancy Nouaimeh
Culture, Leadership and Excellence Consultant.
Shingo Institute Licensed Affiliate. Non-Executive Board
Member (ASQ -2019-2024-, and ISCM)

THE EXPERIENCE

Having lived and worked in Dubai and Saudi Arabia for over 20 years, two of the most culturally diverse and dynamic regions in the Middle East, I have observed numerous inhibitors to project success, many of which are deeply tied to cultural dynamics. In workplaces comprising employees from 30, 40, or even 50 different nationalities, challenges are significantly amplified. Furthermore, some challenges, though not traditionally attributed to cultural differences, are undeniably heightened in such environments.
These include:

- *Communication Gaps*. Differences in language fluency, professional terminology, and communication styles across cultures often lead to misunderstandings and misalignment. This is particularly evident in regions where English may be the common business language, but fluency and interpretation vary widely. The results are often miscommunication or lack of communication which could delay decisions and hence impact project delivery.
- *Cultural Expectations*. Varying attitudes toward hierarchy, decision-making, and time management are especially pronounced in environments like Dubai and Saudi Arabia, where traditional values often intersect with global business practices, sometimes creating friction in stakeholder alignment.
- *Resistance to Change.* Cultural differences can intensify perceived threats during periods of change, increasing resistance among team members which delay projects.
- *Alignment with Objectives*. Ensuring all stakeholders—across a mix of expatriate and local teams—understand and commit to project goals can be challenging, given the diversity in expectations, priorities, and working styles.

DOI: 10.1201/9781003544807-41

- *Short-Term Engagement.* The short-term nature of many expatriate employees' contracts can sometimes lead to unhealthy competition on projects. If the organization's culture is not strong enough to provide a protective and inclusive environment, this competition can undermine collaboration and project success.

Other issues that are also encountered on projects, which are particular to this region but not necessarily linked to cultural aspects are as follows:

- *Project Scope Creep.* Projects face many changing requirements and poorly defined scopes, with a tendency to overpromise, leading to scope creep, which impacts timelines and budgets.
- *Time Pressure.* The Middle East is known for ambitious projects with tight deadlines, creating pressure that can lead to compromised quality, burnout, and rushed decisions.
- *Dependency on External Consultants.* While consultants bring expertise, overreliance on them, as is often seen in the region, brings external input without building internal capability, which can limit the sustainability of project outcomes.

Here is what I do to tackle such issues:

- To bridge communication gaps, I prioritize cross-cultural communication training for all team members. This helps them understand the nuances of language, professional jargon, and different communication styles. In addition, I ensure that a common language (usually English) is used for business communication, and where necessary, provide translation services or language support. I also encourage the use of visual aids, which can help bridge language barriers and improve understanding across teams.
- I also address differing cultural expectations by conducting cultural sensitivity workshops to raise awareness about varying attitudes toward hierarchy, decision-making, and time management. I ensure that leadership aligns with both local customs and global business practices, modeling behavior that respects cultural norms.
- To address resistance to change, I involve stakeholders early in the change process, positioning them as champions to drive acceptance. I implement incremental changes, respect cultural sensitivities, and clearly communicate the benefits of change to align with organizational and cultural values.
- I organize cross-cultural workshops and meetings where stakeholders can align on project goals and discuss expectations. I focus on building teamwork and trust, ensuring that the emphasis is on collective success rather than individual achievements.
- To address scope creep, I work closely with stakeholders to ensure that promises made align with the project's capabilities. Regular reviews and

updates help manage evolving needs and prevent scope creep from impacting timelines or budgets.

- As a consultant, I promote knowledge transfer and encourage internal teams to take ownership of project outcomes, and I focus on developing internal capabilities, leading to more sustainable and self-sufficient project management practices.

My suggestions for those leading, developing, or planning projects in organizations **where culture may be an issue** are as follows:

1. Take time to understand the existing culture and the systems in place that may be driving negative behaviors.
2. Acknowledge cultural issues and ensure that people are aware of these challenges. Ignoring them won't make them disappear.
3. Address cultural issues as part of the project planning process and integrate them with change management strategies to improve the chances of project success.
4. Continuously assess the cultural dynamics and be willing to adapt project plans and leadership approaches based on feedback from the team. This flexibility helps ensure that cultural challenges are addressed in real time, leading to more successful outcomes.

Green Belts, Black Belts, Lean Practitioners, and Agile Scrum Masters are often well-prepared to face certain aspects of resistance to change, but they may not always be fully equipped to handle cultural issues unless they have specific training or experience in that area.

Here's a breakdown why they may not be fully prepared:

- *Focus on Technical Skills.* Many certifications and training programs for Lean and Agile practitioners emphasize tools, methodologies, and process improvement techniques (e.g., DMAIC, Kaizen, Scrum frameworks) rather than the human or cultural elements involved in change.
- *Limited Cultural Awareness.* Resistance to change is not only about process but also about people. Understanding and managing cultural differences, interpersonal dynamics, and emotional responses to change often requires additional skills that go beyond Lean or Agile practices.
- *Lack of Change Management Training.* While Lean and Agile practitioners are excellent at driving process improvements, change management frameworks that focus on people, culture, and overcoming resistance are often not deeply integrated into their primary training.

The skills that they would need most are:

- *Emotional Intelligence (EQ).* The ability to understand and manage one's own emotions, as well as the emotions of others, is key to addressing resistance to change. Practitioners who possess high EQ can navigate difficult conversations, manage conflicts, and build trust, all of which are essential when handling cultural resistance.

- *Change Management Skills.* While Lean and Agile provide the tools for improving processes, effective change management skills are crucial when dealing with resistance, especially cultural resistance.
- *Cultural Sensitivity and Awareness.* Being able to recognize and respect different cultural norms and values can play a huge role in managing resistance to change. Practitioners who understand the cultural context in which they work can adjust their approach to be more inclusive and empathetic, improving acceptance and collaboration.
- *Collaboration and Facilitation Skills.* Building consensus and engaging people from diverse backgrounds can help reduce resistance. Practitioners with strong facilitation skills can encourage open dialogue, address concerns, and create a sense of shared ownership, which is vital for overcoming resistance to change.

Note by the editor: Although your personnel may not be from different countries, they might have varying ages, educational backgrounds, and work experiences, and may have different expectations regarding their future.

33 Working with an Open Mind for Change in the Automotive Industry

Hernán Eduardo Galdeano
Former Director of the After-Sales and Customer Service
Division for Ford Argentina and the South America Group

THE ORGANIZATION

At the time of this experience, this leading automotive company had approximately 3,000 employees, of whom 120 worked in its after-sales division. In this organization, my responsibilities included service engineering, warranties and goodwill, extended warranties, development, validation, procurement, logistics, and parts and accessories sales. In addition, I was responsible for training the dealer network and customer contact center, on-route assistance, and customer satisfaction survey management. I reported internally to the president of the local branch and was matrixed to a LATAM director and the global VP of after-sales at the head office in the USA.

PERFORMANCE COMPARISON WITH OTHER MARKETS

The organization was managed with a Business Operational System that included certain KPIs, and operations appeared to be running smoothly. Furthermore, the company had established a set of procedures that encompassed the majority of its operational scope. However, there were some areas where the values of these KPIs could be enhanced in comparison to those of other markets in Europe and the USA.

BACKGROUND: FIRST IMPRESSIONS

I had previously served as a BoD member and also had several years of experience in customer services. There were some changes, including rotations and the incorporation of new members with new responsibilities, compared to my previous experience.

DOI: 10.1201/9781003544807-42

As I gained a deeper understanding of the people and processes, I recognized that we had a significant opportunity for improvement, with the potential to make further changes.

I had always believed, or at least had the intuition, that any project on technical aspects that was intended to be effective, transformative, lasting, and attractive for all parties involved should have its own intrinsic energy, capable of sustaining it and driving continuous improvement beyond the initial external energy that would set it in motion.

THE GREAT CHALLENGE

I knew that this "own energy" would have its power plant in the sociocultural aspects, which are closely linked to each project and have an impact on the organizational culture of the area that has to carry it out.

In short, from now on, **the key would be to work much more closely with people** (I never liked the term "Human Resources"), beyond relying on their talents, to bring out their abilities and encourage creativity without fear of making mistakes.

RESTARTING

- We redefined our quality policy with a different vision: we would no longer be satisfied with simply: "Satisfy the customer and ensure their mobility" but, inspired by Karl Albrecht's customer value hierarchy, we would seek to **"Achieve the customer's emotional commitment to our brand, through a great after-sales experience."**
- We redefined the key objectives for the business plan and strategy, so that our mission will lead us to be the best, in a sustainable way, in five years, or sooner.
- We redefined the management structures, the necessary profiles and elaborated a schedule of training plans to fill the gaps and thus fulfill the mission that would lead us to materialize the vision.
- We redesigned a specific quality management system for the after-sales division, which was certified by a third party under ISO9001 in 2009.
- With these actions underway, we invited our dealers to join us and follow our path, since they were, by definition, "the closest to the customer."
- The entire network started working with BOS and KPIs that were aligned and complementary to those of the OEM.
- Every Thursday of the 50 weeks of the year my staff and I visited dealers and customers, to ensure the necessary "intimacy" and to measure their progress.
- Once a week my staff and I had a meeting of no more than one hour, in which we went through the BOS and its KPIs, to make sure that its data would allow us to manage properly in the short and long term.
- At all times my office operated on an "open door" basis and every member of my office had the opportunity to have one-on-one "skip level" meetings with management.

- We fostered a program of knowledge of the functions of each member of the division with an innovative plan called "Partners for One Day," which strengthened us as a team.
- Dealers received technical, commercial, and management training, developed through agreements with universities and private tertiary institutes, through a program called **"The Customer Is Our Boss,"** shared with marketing and sales management. It included a successful annual competition with important prizes for those dealers who met or exceeded their objectives, especially those with "high customer satisfaction."
- We created a special incentive for those dealers who develop a quality management system to improve their business management and have it certified by a second party.

RESULTS

After five years of work, the management results showed the following:

- Customer satisfaction with after-sales service doubled. I clarify that the measurement was very demanding because it was a polynomial expression that counted satisfied and very satisfied customers, while subtracting points in the categories "partially dissatisfied" and "totally dissatisfied."
- The "pulse survey" of the customer service department was the highest in the Branch in four out of five years, with very high values and several percentage points higher than the rest of the offices.
- The Profit Before Taxes (PBT) contribution of the after-sales division to the total PBT of the branch increased by more than 500% in constant currency over five years.
- The "customer permanence with dealers" period, which extends beyond the end of the warranty period, increased by approximately 30%.
- For the first time in years, a record number of people from other parts of the company wanted to join the after-sales and customer service division.
- Both our team and our dealer network understood more clearly each year that "The Customer Is Our Boss."

IT HAS BEEN ALMOST 10 YEARS SINCE I LEFT THE COMPANY

I see it more clearly: None of this great experience would have been possible without a great team and the support of the company's leadership, who believed in this change. And even with their support, none of this would have been possible if we had approached it as one or more individual projects, with technical or operational aspects, disconnected from each other and far from the social and human aspects. I would like to highlight the support of the HR department.

By working with an open mind for change, without fear of the freedom to take risks and even make mistakes, we created a new organizational culture in which motivation and people management achieved transcendent goals.

34 Movistar's Experience in Argentina

Luciana Barrera
Director of Digital Transformation
at Movistar Argentina

THE ORGANIZATION

Telecommunications company with approximately 15,000 employees and nation-wide operations.

THE PROJECT—DIGITALIZATION

High strategic and operational impact, long term, and with the following objectives:

- Simplify IT systems. Design, implement, and support digitalization to go from operating with more than 200 different systems that supported the end-to-end customer experience to a Full Stack system. The project included connecting the new system and retiring all previous systems.
- Manage the project in a culturally complex environment. The organization is multicultural. Part of the workforce has experience in fixed telephony and part has experience in mobile telephony as a result of the merger of two competitors. Part of the employees—the majority—have been part of the company under state management and the other part of the employees have experience only in the privatized organization.

VISION TO BE ACHIEVED—THE CHANGES

- Simplification and digitalization of processes to enhance the customer experience and satisfaction.
- Implications in terms of: (i) change of tasks—elimination and changes; (ii) change of structure—reduction of management levels and change in the organization; and (iii) significant reduction of staff (throughout the organization—except for 50 people in one unit).

DOI: 10.1201/9781003544807-43

BACKGROUND

- Several instances were identified, all with the same result: the previous ones had not been closed, which implied the continuity of those to be replaced.

THE BIG CHALLENGE—CULTURAL

The project faced significant operational challenges, with cultural development emerging as a crucial area that required particular attention.

- Cultural, rather than technological: How to motivate and engage employees in the project to achieve a transformative future without the limitations of the past? Especially when many would experience changes in their roles and even job security, how to align the diverse experiences and habits—managerial and operational—among the staff?

PROJECT DEVELOPMENT

The company recognized the necessity to address two key areas: operational and cultural.

- The operational aspect was overseen by the IT director and a team of approximately 100 individuals.
- The cultural aspect was overseen by the transformation director and the HR director. A change management group was established for the project, comprising three individuals. The processes management team was also incorporated into this area.
- Initially, both sides had their own plan, but there was no unified, integrated plan.

PROBLEMS FACED

As previously noted, the project encountered several challenges along the way. Luciana's key takeaways were as follows:

- The need to combine, in project teams, people with enormous experience and knowledge of the organization and technology, with people with a vision of the future but limited knowledge of the organization and technology. The "battle" between the past and the future.
- Conditionality of the designs and solutions due to resistance originated in the loss of power and work. And with the union favoring solutions that would minimize the loss of capital.
- There was a perception of winners and losers.

IMPACT ON THE PROJECT

Cultural challenges affected the project in the following ways:

- The potential impact on project quality due to the risk of losing future differentiation by prioritizing the voice of experience or losing the voice of experience by idealizing the future.
- The initial mistake was not to listen with generosity to those with experience.

SOLUTIONS

The most outstanding solutions were as follows:

- Bring to the table people with both characteristics—knowledge and vision of the future. Install analysis and decision-making mechanisms based on "understanding the cause" of each party's rationale—understanding the position of others. Turn discussions into listening and decision-making mechanisms to manage the tension between the two groups and reach consensus.
- Level knowledge.
- Give everyone a seat at the table.
- Stage project design.
- (i) A first project phase to define the basic technical aspects and the final migration. A second phase to achieve operational and customer buy-in.
- (ii) In the second phase, and due to internal moves of the technical leader to Spain, the "cultural" leader took over the systems management to manage the change from "inside" the project.
- (iii) In a third phase, that of migration, the shutdown and "reinvention" of people for relocation. And again, a change in the organization: leadership moves to lead from processes, experiences, and change management.
- Place strong emphasis on listening and valuing contributions to define the new system.
- Retrain the old, those with organizational and operational knowledge: 85% of BI staff came from the business and were retrained with analytical skills.
- Negotiate with the union; included in the decision-making table, with continuous anticipation of steps to be taken.
- Carry out voluntary retirements. Back-Office reduced from 2,200 to 200 people.
- Redesign the structure in line with the project—see project phases.

LEARNING

The main lessons learned were as follows:

- There was a lack of plans to address contingencies and technical contingencies. The emphasis placed on the "ideal world" meant that there was a lack of capacity to face problems and contingencies.
- There was a valorization of the contribution made by those involved in planning and executing change management. Today, the structure is decentralized and agile.

Note by the editor: Both cases show (i) Change management applies not only to the organization's personnel but to everyone in the value chain. (ii) The significance of support from top executives is clearly conveyed. (iii) The importance of collaborating with other functions within the organization is also evident. (iv) The culture and its impact on the projects should not be underestimated. (Chapters 33 and 34)

References

Argyris, Chris. *Reasoning, Learning and Action*. Jossey-Bass. (1982).

ASQExcellence. *IoE Automotive Industry Report—Insights on Excellence*. ASQExcellence. (2021).

Behar, Howard and Goldstein, Janet. *It's Not about the Coffee*. Penguin. (2009).

Bertin, Marcos and Watson, Gregory. *Corporate Governance, Quality at the Top*. Goal/QPC. (2007).

Boston Consulting Group. American Society for Quality and the Deutsche Geselischaft fur Qualitat. *Quality 4.0* – https://asq.org/quality-resources/quality-4-0. (2023).

Bladek, Oliver, Deighton, James, Dunn, Alison, Huizenga, Tip, and Walden, Wesley. *The Wisdom of Transformations: How Successful CEOs Think about Change*. McKinsey. (2019).

Breyfogle III, Forrest W. *Implementing Six Sigma*. John Wiley & Sons. (2003).

Bridger, Emma and Gannaway, Belinda. *Employee Experience by Design, How to Create an Effective EX for Competitive Advantage*. Kogan Page. (2021).

Cameron, Julia. *The Artist's Way*. Editorial Tquel. (2007).

Cecchi, Oscar, Molteni, Raúl, Zapiola, Ernesto. *Business Strategies*. Market. (2000).

Change Management Learning Center. *The Psychology of Change: Understanding the Guiding Principles of Effective Change Management*. Prosci. (2025).

Clatworthy, Simon. *The Experience-Centric Organization, How to Win Through Customer Experience*. O'Reilly. (2019).

Covey, Stephen, Merrill, Roger, A. and Merrill, Rebecca R. *First Things First*. Simon & Schuster. (1995).

De Bono, Edward. *De Bono's Thinking Course*. Plaza & Janes. (1990).

De Smet, Aaron, Schaninger, Bill, Smith, Matthew. *The Organizational Health and How to Capture It*. McKinsey. (2014).

DEC Association. *The Profitable Customer Experience, a Handbook for Managers and Professionals*. DEC Association. (2017).

Deming, William E. *Out of the Crisis*. Massachusetts Institute of Technology. (1986).

Disney Imagineers. *The Imagineering Way, Ideas to Ignite Your Creativity*. Disney Editions. (2003).

Disney Imagineers. *The Imagineering Workout*. Disney Editions. (2005).

Echeverría, Rafael. *Ontology of Language*. Comunicaciones Noreste. (2003).

Generación Mas, *Guía de Estudio - Aprendizaje*. Generación Mas. www.generacionmas.com (2023).

Goncalves, Vicente and Campos, Carla. *The Human Change Management Body of Knowledge*. Human Change Management Institute. (2013).

Guaspari, John. *I Know It When I See It*. American Management Association. (1985).

Gurr, Bob. "Original Imagineer." *The Imagineering Way*. Disney Editions. (2003).

Hammond, Kenneth. *The Psychology of Egon Brunswick*. Holt, Rinehart and Winston. (1966).

Harrington, H. James and Bertin, Marcos. *Corporate Governance, for Small to Mid-Sized Organizations*. Patton Professional. (2009).

Hiatt, Jeffrey M. *ADKAR, How to Implement Successful Change in Our Personal Lives and Professional Careers*. Prosci Research. (1967).

International Academy for Quality. *Quality in Governance Think Tank, Quality at the Top, a Quality Guide for Boards.* (2023).

Isaacson, Walter. *Elon Musk.* Debate, 2024.

Jaques, Elliott. *Requisite Organization, the CEO's Guide to Creative Structure & Leadership.* Cason Hall. (1992).

Jekiel, Cheryl. *Lean Human Resources, Redesigning HR Processes for a Culture of Continuous Improvement.* CRC Press. (2011).

Juran, Joseph M. and Godfrey, Blanton A. *Juran's Quality Handbook.* McGraw-Hill. (1999).

Kotter, John P. *Leading Change.* Harvard Business Review Press. (2012).

Kotter, John P. *XLR8 Accelerate.* Harvard Business Review Press. (2014).

Liedtka, Jeanne and Ogilvie, Tim. *Designing for Growth, a Design Thinking Tool Kit for Managers.* Columbia Business School. (2011).

Liker, Jeffrey K. and Meier, David. *Toyota Way Fieldbook, a Practical Guide for Implementing Toyota's 4Ps.* McGraw-Hill. (2006).

Lippitt, Ronald, Watson, Jeanne, and Westley, Bruce. *The Dynamics of Planned Change.* Harcourt, Brace and Company. (1958).

McKinsey. *Innovation in a Crisis: Why It Is More Critical Than Ever.* McKinsey. (2020).

Molteni, Raúl. *Change Management Is Not a Fad It's a Fact-Based Need.* Sandholm Institute 50th Anniversary Smogasbord. (2022).

Molteni, Raúl. *Creating Organizational Competencies for Transformation.* CIDEH, Peru. (2021).

Molteni, Raúl. *Embracing Change.* Asociación Popular de Ahorros y Préstamos. Dominican Republic. (2021).

Molteni, Raúl. *How Social Matters Impact on Results.* Molteni Consulting Group. (2014).

Molteni Raúl. *Transformation and Digitalization Require a Deep Understanding of Quality and Culture.* Molteni Consulting Group. (2023).

Molteni, Raúl and Cecchi, Oscar. *The Lean Six Sigma Leadership,* 2nd *ed.* Ediciones Macchi. (2005).

Østergaard, Kris. *How Big Companies Can Simultaneously Run and Reinvent Their Businesses.* Singularity Hub. (2019).

Palmer, Brien. *Making Change Work, Practical Tools for Overcoming Human Resistance to Change.* ASQ Quality Press. (2004).

Paulise, Luciana. *We Culture, 12 Skills for Growing Teams in the Future of Work.* Quality Press. (2022).

Prosci Best Practices in Change Management, Benchmarking Report. Prosci. (2016).

Quality 4.0. Boston Consulting Group, American Society for Quality and the Deutsche Geselischaft fur Qualitat. (2019).

Quinn, Robert. *Wisdom for Change.* Prentice Hall. (2015).

Ramanathan, Narayanan and Vandenbrande, Willy. *How Companies Can Apply Quality to Address Planet Earth Concerns. Quality in Planet Earth Concerns Think Tank.* International Academy for Quality. (2019).

Ramanathan, Narayanan and Watson, Gregory H. *Quality Manifesto for the 21st Century.* International Academy for Quality. (2021).

Rother, Mike. *Toyota Kata, Managing People for Improvement, Adaptiveness, and Superior Results.* McGraw-Hill. (2010).

Sadler, Philip. *Designing Organizations, the Foundation for Excellence.* Kogan Page. (1998).

Schein, Edgar H. *Culture and Leadership.* Jossey-Bass. (2010).

Senge, Peter M., Peter Roberts, Ross, Charlotte, Smith, Richard, and Art, Bryan Kleiner. *La Quinta Disciplina en la Práctica*. Granica. (1995).

Sheenan, Jeff. *Customer Experience Management, Field Manual*. Boston Business Books. (2019).

Shtogren, John. *Models for Management, the Structure of Competence*. Teleometrics Int'l. (1990).

Sinek, Simon. *Start with Why*. Portfolio. (2009).

Stanier, Michael Bungay. *The Coaching Habit, Say Less, Ask More & Change the Way You Lead Forever*. Crayons Press. (2016).

Taylor, Frederick Winslow. *The Principles of Scientific Management*. Dover Publications. (1997).   www.amazon.com/Principles-Scientific-Management-Frederick-Winslow/dp/0486299880

United Nations. *17 Goals to Transform Our World*. United Nations. www.un.org/sustainable development. (2023).

Vandenbrande, Willy. *Reflections on Quality in 10½ Columns*. MakeWay Books. (2024).

Index

Note: Endnotes are indicated by the page number followed by "n" and the note number e.g., 111n5 refers to note 5 on page 111.